ESSENTIALS OF DEVELOPMENTAL PLANT ANATOMY

Essentials of Developmental Plant Anatomy

Taylor A. Steeves

Vipen K. Sawhney

UNIVERSITY PRESS

Oxford University Press is a department of the University of Oxford. It furthers
the University's objective of excellence in research, scholarship, and education
by publishing worldwide. Oxford is a registered trade mark of Oxford University
Press in the UK and certain other countries.

Published in the United States of America by Oxford University Press
198 Madison Avenue, New York, NY 10016, United States of America.

© Oxford University Press 2017

All rights reserved. No part of this publication may be reproduced, stored in
a retrieval system, or transmitted, in any form or by any means, without the
prior permission in writing of Oxford University Press, or as expressly permitted
by law, by license, or under terms agreed with the appropriate reproduction
rights organization. Inquiries concerning reproduction outside the scope of the
above should be sent to the Rights Department, Oxford University Press, at the
address above.

You must not circulate this work in any other form
and you must impose this same condition on any acquirer.

CIP data is on file at the Library of Congress
ISBN 978-0-19-065705-5

9 8 7 6 5 4 3 2 1

Printed by Sheridan Books, Inc., United States of America

*Dedicated to the memory of the Late Taylor Steeves (co-author of this book).
A distinguished scholar, teacher, gentleman, mentor, and dear friend.*

VKS

Contents

Preface ix

1. *Introduction* 1

2. *The Plant Cell* 7
    The Protoplast 8
    The Cell Wall 13
    Cell Growth 16

3. *The Flower* 17
    Inflorescence Types 19
    Flower Morphology 19
    Variations in Flower Morphology 22
    The Induction of Flowering 27
    Flower Development 29

4. *Reproduction* 31
    Vegetative Reproduction 32
    Sexual Reproduction 34
    Overview of Sexual Reproduction 42
    Apomixis 43

5. *Embryo, Seed, and Fruit Development* 44
    Patterns of Embryo Development 44
    Somatic Embryogenesis 51
    Seed Development 52
    Fruit Development 55

6. *Shoot Morphology and Development* 57
   - Growth of the Shoot 59
   - Shoot Apex and the Shoot Apical Meristem 60
   - Shoot Expansion 63
   - Shoot Branching 65
   - Shoot Modifications 66
   - Reproductive Shoots 67

7. *Plant Cells and Tissues* 69
   - Cell Types and Tissues 70

8. *Tissues of the Stem* 81
   - The Dicotyledonous Stem 82
   - Three-Dimensional Organization of the Vascular System 84
   - Differentiation of Tissues in the Stem 86
   - The Monocotyledonous Stem 91
   - The Stele 92

9. *The Leaf* 95
   - Leaf Form 96
   - Tissues of the Leaf 101
   - Variations Related to the Environment 104
   - Leaf Development 106

10. *The Root* 111
    - Root Systems 112
    - Root Associations 113
    - Root Apex and the Root Apical Meristem 115
    - Tissues of the Root 118
    - Root Branching 123
    - Shoot Buds from Roots 125

11. *The Secondary Body* 127
    - The Vascular Cambium 128
    - Secondary Xylem (The Wood) 133
    - Secondary Phloem 136
    - Secondary Growth in Monocotyledons 138
    - Periderm 139

GLOSSARY 143
BIBLIOGRAPHY 155
INDEX 157

Preface

THE SEED TO put together this book was sown several years ago by the late Professor Taylor Steeves and myself with the objective of providing, for a growing community of students and researchers in plant developmental genetics and molecular biology, an abridged version of plant anatomy. At the time of our proposal to Oxford University Press, there were excellent plant anatomy texts available, including *The Anatomy of Seed Plants* by Katherine Esau, *Plant Anatomy* by Abraham Fahn, *Plant Anatomy* (Parts 1 and 2) by Elizabeth Cutter, and *Plant Anatomy* by James Mauseth, and more recently other books such as *Esau's Plant Anatomy* by Ray Evert and *Plant Anatomy: An Applied Approach* by David F. Cutler, Ted Botha, and Dennis Wm. Stevenson. Although most of these books in plant anatomy are superb in their content, they are exhaustive in detail. By using several examples from different groups of plants, they provide a comparative account of plant structure and function as well as evolutionary trends in plant structure. We felt then, and even now, a need for an abbreviated text in anatomy with a plant developmental perspective. This book should be useful to researchers and students in plant biology in general, and in developmental genetics and molecular biology in particular. This book could also serve as a complementary text to an excellent laboratory exercises volume, *Teaching Plant Anatomy* by R. Larry Peterson, Carol Peterson, and Lewis H. Melville, published in 2008.

It has taken a long time to bring this project to fruition, and there are several reasons for this. Initially, Prof. Steeves and I tried to fit the writing work in between our other commitments of teaching, research, and administrative work (both Prof. Steeves and I served as Head of the Biology Department at different times in our university). After we had a draft of the text, we exchanged our writings for comments on the content and uniformity in writing style. We then sought input from Prof. Ian Sussex of Yale University, a world-known developmental botanist, who provided valuable comments and suggestions. Then, while we were working on getting suitable illustrations for the book, Prof. Steeves took ill. That was followed by his prolonged illness, and then his passing in 2011, contributing to further delays. After my retirement in the summer of 2014, I took it upon myself to complete this project, especially in honor of my late colleague and dear friend, Taylor Steeves.

There are several people who deserve a big thanks for the completion of this project: Prof. Ian Sussex for his thorough reading of the text and for his constructive comments (unfortunately Prof. Sussex passed away earlier this year), Prof. Ed Yeung for his comments on the text and for providing a number of excellent images, Dr. Margaret (Peggy) Steeves for editing the entire text, Marlynn Mierau for his very competent and enormous help in putting together the illustrations, and Profs. Art Davis and Ken Wilson for their comments on some chapters. Finally, I am indebted to Jeremy Lewis and the Oxford University Press for their patience and support of this project, and to David Joseph and the production department of Newgen KnowledgeWorks for their competent and efficient work for the completion of this project.

Vipen Sawhney
University of Saskatchewan
Saskatoon

1

Introduction

ANATOMY IS THE science that deals with the structure of organisms. The word itself comes from the Greek *anatom*, "a cutting up," and implies that the analysis of structure is more profound than a simple external study. In plant science, anatomy usually indicates a consideration of cell and tissue organization, often by the cutting and microscopic examination of sections, rather than the observation of general form. Nevertheless, the external form of a plant cannot be ignored if the cell and tissue organization is to be interpreted meaningfully.

A sound knowledge of structure provides an essential foundation for many other aspects of plant biology. The understanding of function depends heavily upon a clear picture of the structural components that perform the function. Comparative anatomical data provide some of the most useful guides to taxonomic relationships and evolutionary derivations. Plants have developed many structural modifications in relation to particular environmental conditions, and their investigation contributes to the understanding of environmental adaptations. In many instances anatomical information is required for utilitarian purposes such as the characteristics of material like wood or fibers, the textural quality of food products, or the ability of certain plants to resist insect or disease attack. A great deal of modern anatomical work is inspired by the need to understand how plants grow and develop and consists of the analysis of the progressive elaboration of the organs and tissues of which the plant is composed. The current application of molecular genetics to the study of plant development emphasizes the importance of a sound understanding of plant structure. Such information is vital for an understanding of the genetic and molecular control of plant growth and function. This text will present a brief account of the essential features of plant anatomy rather than an exhaustive treatment of the subject. It is intended to provide the basic information required by students in their study of a variety of other aspects of plant biology.

In theory a text on plant anatomy could deal with the entire plant kingdom, but there are good reasons why this is not ordinarily the case and will not be done here. The major difficulty of so broad a coverage is the immense difference in organization between, for example, the simple plants that we call algae and the structurally complex flowering plants or angiosperms. Consequently this text, like most in plant anatomy, will deal only with the vascular plants, sometimes grouped together in the single Division Tracheophyta and sometimes separated into several divisions. Within this assemblage the major emphasis will be on the seed-bearing plants and most particularly on the angiosperms. The emphasis on flowering plants is justified by the dominant role played by these plants in the modern terrestrial vegetation of the earth and their overwhelming economic significance. This limitation still leaves a wide range of structural variation, but there is an overall plan of organization that is general.

A common and distinguishing feature of all of the vascular plants is the occurrence of a specialized internal conducting system consisting of two tissues. Water and dissolved mineral elements are transported in the xylem while organic substances (e.g., sugars and hormones) and some minerals are translocated by the phloem. These two components, collectively known as vascular tissue, occur in a constant relationship and extend to all parts of the plant body. Although plants of other groups may have mechanisms of long-distance transport, none have these particular tissues. Thus the designation "vascular plants" is an appropriate one.

In addition to this diagnostic feature, there is an organizational plan that is characteristic of the vascular plants. The plant body is composed of two systems, the shoot system and the root system (Fig. 1.1). The shoot system consists of an axis or stem that bears a succession of lateral organs or leaves. Typically the shoot is upright in orientation and radial in symmetry, but horizontal shoots are not rare and those with bilateral symmetry are known. The shoot system may consist of a single leaf-bearing axis but more commonly it is branched, often extensively. The shoot system also produces the reproductive structures of the vascular plant, and both flowers and cones are considered to be modified shoots. The root system serves to anchor the plant and to absorb water and mineral elements from the soil. Like the shoot system, it is often highly branched.

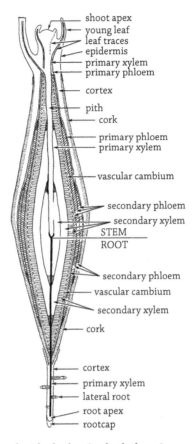

FIGURE 1.1 Diagram of the plant body showing both the primary and secondary tissues produced by apical (shoot and root) and lateral (vascular cambium and cork cambium) meristems.
From *Plant Anatomy* by K. Esau, 2nd edition (1967), reprinted with permission of John Wiley & Sons Inc.

The origin of this body plan is to be found in the early stages of development—that is, in the embryo that develops from the fertilized egg or zygote. In the embryo two centers of continuing growth, called meristems, are set apart early in development. One of these, the shoot apical meristem (SAM), gives rise to the primary shoot, and the other, the root apical meristem (RAM), initiates the primary root. These meristems have the potential for unlimited or indefinite growth and are found at the tip of every shoot and every root in the system. Although the potential for unlimited growth is not always realized, these meristems cause both the shoot and root systems to be open-ended in contrast to the closed system (i.e., limited growth) of the animal body.

The potentially unlimited growth of the shoot and root systems confers certain advantages upon the vascular plant. Because of its cellular structure, the plant cannot carry out the process of cellular replacement or turnover that goes on extensively in the animal body as metabolically active cells wear out. Instead, the plant continually adds on new cells and tissues at its growing tip, as for example in the continued formation of new leaves. Moreover, the plant has no capacity for mobility, but it nonetheless can react to changes in its environment by means of a growth response, as in the case of shoots that grow toward the light or right themselves after being forced into a horizontal position.

The apical meristems of shoot and root have the capacity to produce all of the tissues and organs of a complete plant, and there are many plants that consist entirely of the products of these meristems. The plant body produced by the apical meristems is called the primary body. In many cases, however, the primary body is supplemented by tissues generated by additional meristems that increase the thickness or girth of the plant rather than its linear extension as do the apical meristems. These meristems, the vascular cambium and the cork cambium or phellogen (see Fig. 1.1), cannot produce an entire plant because they give rise only to certain tissues, vascular tissue in the case of the vascular cambium and outer protective tissues in the case of the phellogen. Thus these tissues supplement the primary body and constitute the secondary body of the plant. In plants such as trees, however, the secondary tissues may constitute a large part of the bulk of the body.

The embryo, in which the basic organization of the plant is initiated, results from the process of sexual reproduction. In all vascular plants sexual reproduction occurs in the context of a life cycle and involves the regular alternation of two distinctly different generations (one is sexual and is haploid, and the other is asexual and is diploid), called the alternation of generations (Fig. 1.2). The characteristic vascular plant is the diploid sporophyte while the alternate gametophyte is haploid and is very different (i.e., is much less extensive in its development and is easily overlooked). The best way to introduce this subject is to present an example in which the two generations are easily observed because each has an independent existence. A common fern plant provides such an example.

When the fern plant reaches reproductive maturity, it does not produce flowers and seeds but rather forms spores in sporangia that are distributed in various ways on some or all of the leaves. When these spores are shed and germinate in favorable locations, they give rise not to another vascular fern plant, but rather to a diminutive thallus-like structure that is the gametophyte (see Fig. 1.2). The fern plant is diploid, but meiosis or reduction division occurs in the sporangia so that the spores are haploid, as are the gametophytes that arise from them. The gametophyte produces gametes or sex cells, eggs and sperms, in specialized

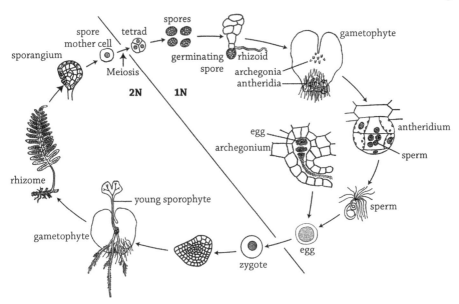

FIGURE 1.2 The life cycle of a typical fern plant illustrating the two alternating generations, the diploid (2N) sporophyte and the haploid (1N) sexual gametophyte.

sex organs or gametangia. Fertilization occurs in the female gametangium or archegonium, and the resulting diploid zygote is retained there, where it begins to form an embryo sporophyte. From this embryo the vascular sporophyte develops, soon becoming the independent fern plant and completing the life cycle.

This life cycle in which two generations, one spore-producing and the other sexual, alternate in regular fashion is characteristic of all vascular plants, but it has many variations. In the ferns and some other lower vascular plants such as club mosses and horsetails, both generations are independent and thus readily recognizable. The seed plants, gymnosperms and angiosperms that make up most of the living vascular plants, have a much less obvious alternation of generations in which the gametophytic generation is drastically reduced and remains dependent upon the sporophytic phase. This topic will be considered in some detail in Chapters 3 and 4.

In plant anatomy, as in many disciplines, there are several different points of view with which the subject may be approached. One widely used approach is the comparative one, in which diverse structural patterns are compared for the purpose of recognizing evolutionary changes and establishing relationships. Alternatively, the approach may be to generalize the structural plan by tracing its development. This developmental approach is more concise and is better suited to a text of this sort. The basic structure of the plant cell, the building block of the plant body, will be presented first. Following this, the treatment will describe the reproductive phase of a flowering plant in the context of the life cycle of alternating phases. It will then trace the development of the embryo leading to the formation of shoot and root systems and both primary and secondary bodies. Thus the structure of the vascular plant will be presented as it actually develops.

2

The Plant Cell

SINCE THE FIRST discovery and naming of cells by Robert Hooke in 1665, knowledge of the structure and function of these entities, and their significance as the basic units of construction of living organisms, has grown steadily. Since the second half of the 20th century, particularly with advancements in microscopy and sophisticated biochemical and molecular techniques, a flood of hitherto unsuspected information has been revealed regarding the internal organization and chemical nature of these units of life. Of necessity this brief survey will touch lightly on this fund of knowledge, stressing primarily those features that are of importance in the study of plant anatomy. Plant cells display such a variety of sizes, shapes, and structures that it is impractical to describe an "average" or a "typical" plant cell. Thus, a general picture of plant cell structure will be presented and variations on this theme will emerge in the subsequent survey of cell types.

A cell is a functional unit of protoplasm, but in plants this unit, the protoplast, is usually enclosed within a rigid cell wall. The protoplast also contains many components that are not protoplasmic. Indeed, the cell wall and the non-protoplasmic components of the protoplast are among the most important cell features for the plant anatomist. The cell wall gives a definite shape to the cell, and these shapes are highly diverse. Plant cells are commonly represented as plane figures in optical sections seen under a compound microscope, but they actually are multifaceted, three-dimensional bodies. The basic shape of cells packed together in a tissue is that of a 14-faced polyhedron, but there are many departures from this shape, particularly cells that elongate during differentiation. The structural components of the plant cell are illustrated in Figure 2.1.

THE PROTOPLAST

Within the cell, the protoplast is bounded by a delicate selectively permeable membrane called the plasma membrane or the plasmalemma. Under the light microscope, it appears as a thin boundary layer, but its structure is revealed in much greater detail by the electron microscope. The plasmalemma is constructed primarily of two phospholipid layers in which protein molecules are both embedded and attached on the surface (Fig. 2.2). The lipid layers have their hydrophilic ends facing outward and their hydrophobic fatty acid tails facing inward. Proteins

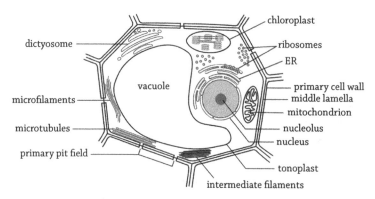

FIGURE 2.1 Generalized diagram of a plant cell showing the cell wall and various organelles in the protoplasm.

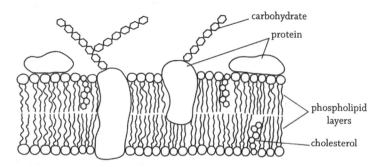

FIGURE 2.2 Diagram of plasmalemma or cell membrane with two phospholipid layers with protein molecules embedded in, and on the outer surface of, the membrane. Some cholesterol molecules are found in the membrane and carbohydrates may be attached to protein molecules.

of a variety of types are present in the membrane, and some large proteins do not fit within the bilipid layers and may project on one or both ends (transmembrane proteins). Carbohydrates are attached to some of the protein molecules (glycoproteins) and to the lipids (glycolipids). There are also some cholesterol molecules dispersed in the lipid layers. This structure of the membrane appears to be characteristic of all the membranes within the cell and is called the unit membrane. The plasmalemma, interestingly, is not a rigid structure but is fluid-like. Indeed, the fluidity of the membrane is important in its selective permeability and the various functions it performs.

The plasmalemma encloses the cytoplasm, which consists of a relatively fluid ground plasm or hyaloplasm within which a number of structural components, or organelles, are included. One of the conspicuous organelles is the endoplasmic reticulum (ER), a system of interconnected tubules or flattened sacs that are bounded by two unit membranes. The ER is a dynamic system involved in the synthesis and transport of both proteins and lipids. Ribosomes, submicroscopic particles of ribonucleic acid (RNA) and protein, are the sites of protein synthesis and are frequently found attached to the outer surface of the ER (rough ER). Other ER units may not have an association with ribosomes (smooth ER), and ribosomes are also found freely floating in the cytoplasm.

The other flattened membranous organelle found in the cytoplasm is the dictyosome or Golgi body. Dictyosomes are small stacks of generally four to seven flattened membranes, called cisternae, and they are involved in the synthesis and secretion of carbohydrates. The carbohydrates, including the cell wall material, are packaged in small vesicles and released at one face of the dictyosome. The ER and dictyosomes are part of the endomembrane system in the cytoplasm, which involves intercommunication between membranes. For example, vesicles released from the ER can travel to and fuse at one end of the dictyosome, and at the other end of dictyosomes vesicles are pinched off and may fuse with other vesicles or with the plasmalemma. Thus, materials synthesized in the ER are transported to dictyosomes and from there released to other parts of the cell or to the outside.

The largest and the most conspicuous structure in the cell is the nucleus. The nucleus houses the bulk of the hereditary information coded in deoxyribonucleic acid (DNA). DNA is associated with basic proteins called histones, and this complex forms fine long strands, the chromatin. Chromatin cannot be seen with the light microscope; however, at the time of mitosis or meiosis the chromatin coils and condenses extensively to form visible structures, the chromosomes. In the nucleus one or more bodies called nucleoli are present, and they have a high concentration of RNA and have a role in the synthesis of ribosomes. The nucleoli and chromatin are bathed in a fluid called the nucleoplasm. The nucleus is bounded by a double membrane, the nuclear envelope, which is not continuous but is perforated by pores, called nuclear pore complexes, through which the nucleus maintains contact with the surrounding cytoplasm. The outer membrane of the nuclear envelope is also often seen to be continuous with the ER, indicating its continuity with the endomembrane system.

Within the cytoplasm two organelles that are large enough to be seen with the light microscope are the mitochondria and the plastids. Both of these organelles are energy-processing entities and are self-perpetuating (i.e., they arise through the division of preexisting organelles), and each has its own DNA and ribosomes. Thus these organelles are capable of synthesizing some of their own proteins. Mitochondria are spherical or more commonly rod-shaped bodies bounded by double unit membranes. The outer membrane is smooth but the inner is infolded to form tubular or plate-like projections called cristae that extend into the interior of the organelle. Mitochondria are the centers of aerobic respiration and the generation of ATP in the cell.

Plastids are a class of several distinct organelles, all of which develop from small precursors called the proplastids present in meristematic cells. Plastids like mitochondria are bounded by a double membrane, the inner one of which produces an internal membrane system. In the chlorophyll-containing plastids called the chloroplasts, the internal membrane system consists of small flattened sacs, the thylakoids, that are stacked one above the other. Each stack of thylakoids, a granum, is connected to other grana by membranes, sometimes called the stroma lamellae. This network of interconnected membranes is suspended in a fluid-like material, the stroma. The pigment chlorophyll is located in the internal membranes, and that is where the light reactions of photosynthesis take place. The carbon fixation reactions of photosynthesis take place in the stroma. Chloroplasts commonly contain starch in the form of granules as the product of photosynthesis; however, in most cases it is transported away for storage elsewhere. Plastids that are colorless and store starch are called amyloplasts. In some cases plastids synthesize oil instead, and these are called elaioplasts. Finally, plant tissues with colors other than green often owe their coloration to plastids called the chromoplasts, which contain yellow, orange, or red pigments of the carotenoid type. The different classes of plastids are interconvertible. For example, in the fall season changes in the leaf color result from the conversion of chloroplasts to chromoplasts. The same phenomenon occurs during fruit ripening. Similarly, the amyloplasts may be converted to chloroplasts upon exposure to light.

The other membranous organelles that are smaller than mitochondria and plastids are peroxisomes and glyoxysomes. Peroxisomes have the specific function of breaking down toxic substances including hydrogen peroxide, produced as a product of some metabolic reactions. Glyoxysomes convert stored lipids into carbohydrates in plants.

The cytoskeleton of the cell consists of a network of non-membranous, filamentous structures that are composed of proteins only. They are microtubules, microfilaments, and intermediate filaments. These organelles are beyond the resolving power of the light microscope. Microtubules are long hollow tubes (25 nm in diameter) and are made up of the protein tubulin, which exists in two forms, α-tubulin and β-tubulin. The major role of microtubules in plant cells is in controlling the orientation of cell wall fibers and thus the cell shape. Microtubules also form and orient the mitotic spindle, an important component of nuclear division. Microfilaments are smaller in diameter (5–7 nm) than microtubules, are long, may occur singly or in bundles, and are made up of the protein actin. Their primary function is in the streaming of the fluid cytoplasm and facilitating the movement and orientation of organelles in the cytoplasm. The intermediate filaments are 10 to 12 nm in diameter, and their major role is to stabilize and anchor the various organelles in a cell.

In most mature plant cells, much of the volume of the protoplast is taken up by a large vacuole or many small vacuoles. The vacuole is bounded by a membrane, the vacuolar membrane or the tonoplast. The contents of the vacuole, vacuolar sap, consist of water with a variety of substances dissolved in it. Indeed, the vacuole is a repository of many of the reserve and waste products of cell metabolism, and its internal osmotic pressure keeps the cell turgid. The phenomenon of wilting is directly related to the excessive withdrawal of water from the vacuoles. The other important role of vacuoles in plant cells is in cell growth. In meristematic cells there are generally many small vacuoles and, as the cell grows, these vacuoles fuse so that at maturity there is usually one large or a few large vacuoles that press the cytoplasm against the containing cell wall. The pressure created by the expanding vacuoles causes the cell to enlarge while the cell wall is in the plastic state.

The non-protoplasmic components of the protoplast that may occur in the vacuole, in the plastids, or freely in the cytoplasm are known as ergastic substances. These substances constitute products of cell metabolism that are reserve nutrients and waste materials, and substances that may serve a protective function against herbivores and pathogens. They often have distinctive structural features that are characteristic of a species and may be useful in plant identification.

The most abundant stored material in plant cells is the carbohydrate starch and, as noted above, it occurs in the form of grains or granules deposited chiefly in amyloplasts but also in chloroplasts. Starch is deposited in superimposed layers around a center or hilum (Fig. 2.3a). Starch grains have a great variety of structural patterns that are highly specific to individual taxa. Less widely distributed but important in some storage tissues, such as seeds, are proteins. These may occur as an amorphous mass or as crystal-like bodies with a specific arrangement. The most distinctive storage protein bodies are the aleurone grains derived from vacuoles and found in many seeds. The aleurone grains have a bounding membrane and may contain an amorphous matrix or various forms of inclusions such as crystalloids or globoids. Following the deposition of the protein in the grain, there is withdrawal of water, resulting in a solid membrane-bound body. Fats and oils collectively known as storage lipids are found mostly in seeds and fruits but are also distributed in other tissues. They may occur in plastids (elaioplasts), as droplets in the cytoplasm, or in membrane-bound bodies called spherosomes. The ergastic substances that do not represent stored nutrients and are most widespread are tannins and crystals. Tannins are a family of phenolic compounds common in plant tissues either in individual tanniferous cells or in systems of such cells. They are located in the vacuole, in the cytoplasm, or in the cell wall. Tannins occur as droplets, granules, or sometimes large amorphous masses. Their function is not clearly understood, but they may deter insect feeding or may offer protection against fungal infection. Recent evidence indicates that some phenolic compounds, particularly flavonoids, also play an important role in the differentiation of some tissues.

Crystals take various forms and commonly are composed of calcium oxalate. Prismatic crystals may occur singly if they are large, or in great numbers if small (see Fig. 2.3b). Spherical aggregates of crystals are known as druses (Fig. 2.3c) and long needle-like crystals, called raphides (Fig. 2.3d), usually form bundles and often nearly fill a cell. Long crystals that occur singly are called styloids, and they may enlarge to the extent of deforming the cell. Crystals are generally found in the vacuole but may also be incorporated in the cell wall. The function of crystals

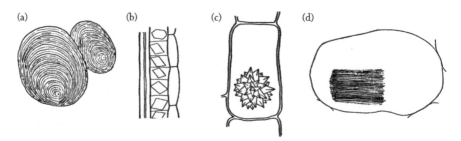

FIGURE 2.3 Diagrams of ergastic substances found in the protoplast: (*a*) starch grains, (*b*) prismatic crystals, (*c*) druses, (*d*) raphides.
From *Plant Anatomy* by K. Esau, 2nd edition (1967), reprinted with permission of John Wiley & Sons Inc.

is not clear, but some of them, particularly raphides, appear to play a protective role by making the tissue unpalatable to herbivores. In addition to calcium oxalate crystals, silica bodies composed of silicon dioxide also occur in plant cells, but less frequently. They resemble crystals but are amorphous in nature.

THE CELL WALL

The cell wall forms an outer jacket of the plant cell (i.e., it is the exoskeleton), and it gives the cell its shape and provides support. In multicellular systems, the cell walls of adjoining cells are attached and held together by the middle lamella. If the cell wall is mechanically removed or digested by enzymes, the cell assumes a spherical shape, indicating the role of the wall in controlling cell shape. Also, when the cell wall is removed and the cell (protoplast) is established in culture, it begins to secrete a new wall, indicating that the cell wall is synthesized by the protoplast. In a growing plant, the cell wall is formed during cytokinesis, the process of cell partitioning after mitosis. Briefly, the process of cell wall formation begins late in anaphase of mitosis as the two sets of daughter chromosomes move to the opposite poles of the cell. With the help of the mitotic spindle, vesicles carrying the cell wall material, and derived from dictyosomes, begin to accumulate at the equator of the cell, where they fuse and deposit the contents to form the cell plate. The expansion of the cell plate is accomplished by a widening ring of microtubules, vesicles, and some ER, known as the phragmoplast or "fence builder" (Fig. 2.4). The phragmoplast expands until the cell plate intersects the parent wall, thereby separating the daughter cells. The first wall that appears in the daughter cells on either side is appropriately designated the primary wall, and the two walls are cemented by the middle lamella (see Fig. 2.1).

The chemical nature of the primary wall has been examined in considerable detail by a number of researchers. The major component of the wall is the carbohydrate cellulose, which consists of long chains of β-1, 4 glucose molecules. Cellulose molecules are bundled together to form microfibrils and are linked to each other by a network of a number of hemicelluloses and pectins that form the wall matrix. The major hemicellulose that interlocks the microfibrils is xyloglucan, and the cellulose–xyloglucan complex forms the framework of the wall in which are embedded pectin molecules. The two major pectins in the primary wall are polygalacturonic acid and rhamnogalacturonans. The distribution of pectins determines the pore size in the wall and therefore regulates the size of substances passing through the wall. The other important component of the primary wall is the protein extensin, which maintains the structure of the wall. The primary cell

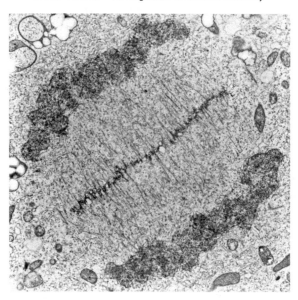

FIGURE 2.4 Electron micrograph showing the phragmoplast with microtubules and Golgi vesicles, and the formation of the cell plate. (Courtesy of Dr. Larry Fowke)

wall is porous and highly hydrated, and water may make up as much as 70% of the wall components. This structure of the wall is found in most flowering plants examined except in grasses, where the cellulose microfibrils are interlocked by a different polysaccharide called glucuronoarabinoxylan. The primary cell wall in grasses also contains some phenolic compounds that bind many of the hemicelluloses to the cellulose fibrils.

The primary wall has no microscopically visible openings in it; however, it is penetrated by cytoplasmic strands, bounded by plasma membrane, which connect the cytoplasm of adjoining cells. These connections, called plasmodesmata (singular, plasmodesma), are initially formed during cytokinesis when the cell plate is developing. As the wall grows, additional plasmodesmata are formed. Plasmodesmata provide the protoplasmic continuum, the symplast, from one cell to the next and are usually concentrated in, but not restricted to, localized regions where the wall is thinner than other parts. These thin areas are referred to as primary pit fields. The cell wall and the middle lamella and intercellular spaces between cells constitute the non-protoplasmic continuum called the apoplast.

All plant cells have a primary wall, but some have an additional wall, the secondary wall, which is laid down interior to the primary wall after the cell has ceased growth (Fig. 2.5).

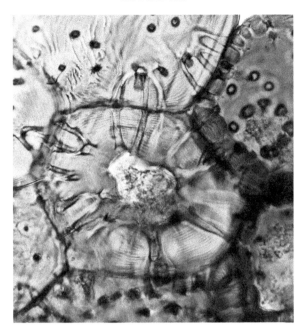

FIGURE 2.5 A sclereid from pear (*Pyrus* sp.) fruit with secondary cell wall. The wall has pits. (Courtesy of Dr. Edward Yeung)

Plant cells that perform the mechanical functions of support (i.e., fibers and sclereids) and the tracheary elements (i.e., tracheids and vessel elements involved in water transport) have a secondary wall. This wall has a high concentration of cellulose, as much as 90% of the wall (e.g., cotton fibers), and is laid down in different layers, each with a definite orientation of microfibrils (see Fig. 2.5). The secondary wall has little or no pectin and very little, if any, of the hemicelluloses that interlock the cellulose fibers in the primary wall. Instead, the secondary wall contains a special substance called lignin, which is a polymer of several aromatic alcohols. Lignin fills the spaces between the cellulose fibers, thus making it a rigid and impermeable wall. The secondary wall, once formed, has no capacity for extension, and most cells with secondary wall are functionally dead when mature. There are no protoplasmic connections through secondary walls, but gaps or pits that expose the primary wall provide connections with adjacent cells (see Fig. 2.5). A pit in one cell commonly faces a corresponding pit in the adjoining cell, constituting a pit pair. In some water-conducting tracheary elements, the secondary wall is not complete but is laid down in the form of rings or in helical (spiral) form so that the cell, even though dead, may undergo some passive stretching (discussed in Chapter 7).

CELL GROWTH

Plant growth is ultimately dependent on cell growth—that is, cells must enlarge or elongate in order for growth to occur. Because plant cells have a cell wall (the exoskeleton), it must give in or "loosen" for cells to grow. During cell growth, the bonds linking the cellulose microfibrils to xyloglucan are broken, which is accomplished by some proteins and enzymes, including expansins. Thus, the wall, once "loosened" and under turgor pressure built up within the cell, is able to stretch. The microfibrils in the wall are initially oriented horizontally, like the coils in a spring, and there are overlapping coiled rings of microfibrils. During cell growth the microfibrils stretch, like the coils in a Slinky toy as it goes down the steps, or the stretching of coils of a book binder, and this stretching contributes to cell elongation. When cell growth has terminated, the cellulose fibrils are held in place by the protein extensin. During cell elongation, additional cellulose fibrils and hemicelluloses are deposited into the wall such that the wall remains of the same thickness as before stretching. Plant cell growth is plastic, not elastic, as the microfibrils after stretching are interlocked by extensin.

3

The Flower

THE FLOWER IS a specialized structure produced primarily for the purpose of sexual reproduction in angiosperms. Therefore, flower formation is essential for both an increase in the population and perpetuation of a species. In addition, since sexual reproduction involves the fusion of male and female gametes, produced after meiosis, flower formation is critical for the introduction of genetic diversity in a species.

The production of flowers is also of considerable importance in the area of agriculture. The formation of fruits and seeds is directly dependent on flower formation and on sexual reproduction, with few exceptions (discussed later). Fruits and seeds develop from different parts of the female reproductive organ, fruits from the ovary and seeds from the ovules present in the ovary. Seeds, especially certain cereals and legumes, are a major source of food for a large portion of the human population. In addition, fleshy fruits of various types are important foods for people around the world and for many animals. Finally, the aesthetic value of flowers and their contributions to the field of floriculture is another important role of flowers in our world.

The pattern of flower formation varies in angiosperms. In annuals, plants that grow for one season, the vegetative growth of the plant terminates with flower formation. Biennials have a two-year lifespan, remaining vegetative in the first year and flowering in the second. In perennials, flower formation is a recurring event and occurs in each growing season. In perennial plants, such as cherry and apple fruit trees, flowering typically occurs annually.

The position of flower formation on a plant varies among species. Flowers may be produced singly or as a group. When produced singly they may occupy a terminal position at the tip of the shoot or in the axil of a leaf, an axillary position. Flowers produced in a group or a cluster collectively constitute an inflorescence. Each inflorescence consists of a stalk on which flowers may be produced along the length of the stalk or at the enlarged shoot tip, and the inflorescence may be branched. There are several types of inflorescences (Fig. 3.1), and these fall into two groups, determinate and indeterminate. In a determinate inflorescence the main axis terminates in a flower and the flowering sequence, from youngest to the oldest, is from the tip down or from the center to the periphery of the shoot tip. In an indeterminate inflorescence the growing point produces lateral flowers

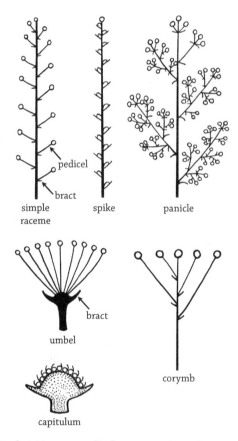

FIGURE 3.1 Diagrams of various types of inflorescences in angiosperms.
From *Botany: An Introduction to Plant Biology* by T. E. Weier, C. R. Stocking, and M. G. Barbour, 5th edition (1974), with permission of Dr. M. G. Barbour.

(on the side of the axis) and the flowering sequence is from the tip of the stalk to the base, and the inflorescence may be branched. A brief description of some inflorescence types follows.

INFLORESCENCE TYPES

A spike has an elongated stalk on which flowers are produced singly and directly on its axis (see Fig. 3.1) as in wheat (*Triticum aestivum*) or barley (*Hordeum vulgare*). The flowers are generally small (called florets), they do not have a stalk (sessile), and they are produced from the base up—that is, older flowers at the base and the younger toward the tip. A raceme is like a spike, but each flower has a stalk or a pedicel, as in mouse ear cress (*Arabidopsis thaliana*) or canola (*Brassica napus*) and other members of Brassicaceae (cabbage family). A raceme may also be branched and is called a panicle (see Fig. 3.1), as in oats (*Avena sativa*). An inflorescence in which flowers are grouped at the tip of a stalk and each flower has a stalk of similar length is called an umbel, as in onion (*Allium cepa*). In a corymb, stalks of unequal length are attached along the axis and bring the flowers to a common level. If the shoot tip is enlarged to produce several flowers, which are crowded together, the inflorescence is called a capitulum or a head (see Fig. 3.1), as in sunflower (*Helianthus annus*). In this case the youngest flowers are in the center and the oldest toward the periphery.

FLOWER MORPHOLOGY

The morphology of angiosperm flowers is highly diverse and is one of the criteria used in the classification system of flowering plants. There is considerable variation between species in terms of the flower form, shape, size, and color, as well as in the number of floral organs and their arrangement in a flower. For example, flowers may be very small and not visible to the naked eye, as the duckweed (*Lemna*) growing in a pond, or they may be nearly a meter in diameter, as the giant *Rafflesia* found in Malaysia. Similarly, the number of organs in a flower may be few to many. There are also some major differences in floral form between the dicotyledons and monocotyledons. The following is a brief account of a generalized flower structure and a description of select examples depicting varied flower morphology.

A flower generally consists of the four basic types of organs attached to the tip of a stalk, the receptacle (Fig. 3.2). Each type of floral organ may be present in one or more whorls. The outermost whorl of organs is the sepals, which are

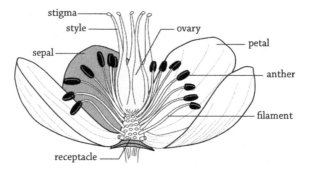

FIGURE 3.2 Diagram of a simple flower showing sepals, petals, stamens, and carpels attached to the receptacle. This is a hypogenous flower.
From *Botany: An Introduction to Plant Biology* by T. E. Weier, C. R. Stocking, and M. G. Barbour, 5th edition (1974), with permission of Dr. M. G. Barbour.

leaf-like structures, usually green, and therefore photosynthetic. Sepals may have hairs (pubescent) on the abaxial (surface away from the shoot tip) or on both the abaxial and adaxial (toward the shoot tip) surfaces. Sepals enclose other floral organs during early stages of flower development and, therefore, have a role in protection of internal organs. All the sepals in a flower are collectively called the calyx.

The organs inner to sepals are the petals. They generally occupy a position alternate to sepals and are usually the colored part of the flower. Petals are also leaf-like structures and may be very elaborate in size and form. They too may be present in one or more whorls. The main function of petals is to attract insects to effect pollination. The petals of a flower are collectively called the corolla. In most dicotyledons, sepals and petals are clearly distinguishable; however, in many monocotyledons and some dicotyledons they are not separable and are called the tepals. The term *perianth* is often used to describe collectively the whorls of both sepals and petals.

Stamens are the male reproductive organs and are positioned inner to petals and usually alternate with them. Stamens may also be present in one or more whorls, and all the stamens in a flower are collectively called the androecium. Each stamen usually consists of a stalk at the base, the filament, which bears an anther at the top (see Fig. 3.2). In some cases, the stamen may lack a filament, or the filament may be very small and inconspicuous. An anther is a lobed structure, usually of two or four lobes, and consists of several specialized tissue layers. Each anther lobe has an outer epidermis, one or two wall layers, a layer of endothecium, and the innermost layer of tapetum, which surrounds the sporogenous tissue (i.e., microspore mother cells or microsporocytes). The endothecium has a role in the opening of an anther for the dispersal of pollen grains at maturity.

The tapetum serves an important function of providing several essential nutrients and metabolites, including certain enzymes, required for the development of microspores and maturation of pollen grains. The sporogenous tissue is located in the center of an anther lobe that after meiosis produces haploid microspores. Each microspore divides by mitosis to form a two- or three-celled pollen grain or the male gametophyte (see Fig. 4.5 in Chapter 4). For details on pollen development see Chapter 4.

The pollen grain is a special structure in seed plants, and its main function is to transport sperm cells (male gametes) to the ovule in an ovary of a carpel, the female reproductive organ. Each pollen grain has a thick outer wall, the exine, that is resistant to decay and protects the sperms from desiccation. The two-celled pollen grains (e.g., tomato) have a large vegetative cell and a small generative cell; the latter divides to form two sperm cells during or after pollen germination. In the three-celled pollen (e.g., *Arabidopsis* and *Brassica* sp.), the two sperm cells are present in the pollen grain at maturity (see Fig. 4.5 in Chapter 4)—that is, before pollen dispersal.

The innermost organ in a flower is the female reproductive organ, the carpel, of which there may be one to many (see Fig. 3.2). Each carpel consists of three parts, the basal swollen ovary region, an elongated style, and a flattened tip, the stigma. In cases where there are several carpels in a flower, they may be separate (apocarpous; e.g., buttercup [*Ranunculus* sp.]) or fused (syncarpous; e.g., tomato or *Arabidopsis*) to form a pistil. All the carpels in a flower make up the gynoecium. The ovary contains one or several ovules that are attached to the ovary wall by a placenta. An ovule consists of one or two outer layers, the integuments, which enclose the inner tissue called nucellus. One cell in the nucellus, the megaspore mother cell or megasporocyte, undergoes meiosis, forming four megaspores; of these, one produces the female gametophyte and the remaining three degenerate. In the majority (approximately 70%) of angiosperms, as in *Polygonum* sp., the female gametophyte consists of seven cells and eight nuclei—that is, an egg cell, two synergid cells, three antipodal cells, and a large central cell with two polar nuclei (see Fig. 4.3 in Chapter 4). For details on the structure and development of the female gametophyte see Chapter 4.

The stigma is the region where pollen grains land and germinate to form pollen tubes. The stigmatic surface is usually made up of elongated papillate cells that secrete various substances that help in both the attachment and germination of pollen grains. After germination, the pollen tube grows through the transmitting tissue of the style to reach the ovary, and then enters the ovule. The length of the style is varied; it can be short (a few millimeters) or several centimeters long, as the silk of corn cob.

In addition to the four classes of organs, other specialized structures may also be found in a flower. For example, nectaries, which secrete sugary nectar that attracts insects and some birds, are formed near the base of many flowers and have an important role in pollination, as in members of the families Brassicaceae and Fabaceae. The number and size of nectaries is varied among species, but they generally possess many stomata near the tip from which the nectar exudes. Another structure found commonly in grasses is the lodicule, which is generally two or three in number in a flower. Lodicules are present toward the base of a flower (see Fig. 3.8 later in the chapter) and are believed to be modified perianth members and have a role in the opening of the flower. Finally, an individual flower or an inflorescence may have leaf-like structures present below called bracts (see Fig. 3.1). If there is a whorl of bracts subtending an inflorescence, it is called an involucre.

VARIATIONS IN FLOWER MORPHOLOGY

As stated earlier, flower morphology is highly varied in angiosperms; therefore, it is impossible to describe a "typical" flower. In addition to variation in the size and number of organs of each type, there are other characteristics, for example the absence of one or more organ types in a flower, the fusion of organs of the same whorl or between organs of different whorls, and the position of the attachment of organs.

Flowers that contain all four classes of organs are called complete. Incomplete flowers, which lack one or more type of organs, are not uncommon in angiosperms. For example, flowers may have sepals but no petals (apetalous; e.g., *Clematis*), or both sepals and petals may be missing, as in the Calla lily. Similarly, one or the other reproductive organ may be missing in a flower. The majority (70%) of angiosperm flowers are bisexual—that is, they contain both the stamens and carpels. However, unisexual flowers containing only stamens (staminate) or only carpels (pistillate) are found in many families. In the family Cucurbitaceae (cucumber and squash), male flowers with stamens are produced at the base of the plant and female flowers toward the tip. Conversely, in corn (*Zea mays*), male flowers are produced at the tip of the plant in the tassel and female flowers in the ear toward the base. Male and female flowers may be produced on the same plant (monoecious), as in corn, or on different plants (dioecious), as in *Cannabis*.

Variation in flower morphology can also result from differences in the point of attachment of floral organs in relation to the gynoecium. Flowers in which sepals, petals, and stamens are attached below the ovary are called hypogenous

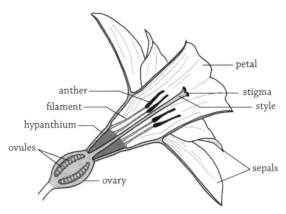

FIGURE 3.3 Diagram of an epigynous flower (e.g., daffodil), in which the lower parts of sepals, petals, and stamens are fused to form a hypanthium, which in this case is fused with the ovary.
From *Botany: An Introduction to Plant Biology* by T. E. Weier, C. R. Stocking, and M. G. Barbour, 5th edition (1974), with permission of Dr. M. G. Barbour.

(see Fig. 3.2), as in *Brassica, Arabidopsis*, lily, or tomato. In these cases, since the ovary is located above all other organs, it is also called a superior ovary. If all the floral organs are attached on top of the ovary, the ovary is epigynous or inferior (Fig. 3.3), as in daffodil or apple. In the epigynous condition the lower parts of sepals, petals, and stamens are fused to form a floral tube called the hypanthium, which may be fused with the ovary. However, in some flowers, as in cherry and plum (members of the family Rosaceae), the hypanthium does not fuse with the ovary and the other organs arise from the edge of a cuplike hypanthium (Fig. 3.4). This condition is named perigynous.

In a flower, individual floral organs may be attached separately to the receptacle so that each organ is distinct. In such cases, the prefix *apo-* is used to describe the condition; for example, aposepaly and apopetaly indicate that all the sepals and petals, respectively, are separate. However, in many flowers, floral organs of one whorl or organs of different whorls may be partially or completely fused. When organs of the same whorl are fused (i.e., are coalesced), the prefix *syn-* is used, as in synsepaly and sympetaly for the fusion of sepals and petals, respectively. Similarly, the fusion of stamens is called synandry and that of carpels syncarpy. Fusion may also occur between organs of different whorls (adnation), for example fusion of stamens to petals as in snapdragon flowers.

The above description is intended to emphasize the immense diversity found in angiosperm flowers. However, within a plant family the flower structure of different species is less variable. The following is a brief account of representative flower structures of some select families.

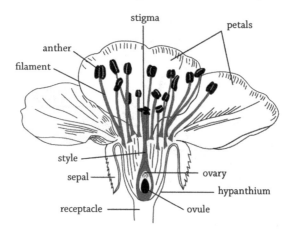

FIGURE 3.4 Diagram of a perigynous flower showing a nonfused hypanthium (e.g., rose flower).

From *Botany: An Introduction to Plant Biology* by T. E. Weier, C. R. Stocking, and M. G. Barbour, 5th edition (1974), with permission of Dr. M. G. Barbour.

RANUNCULACEAE

Flowers of the family Ranunculaceae (buttercup) are characterized by a large number of stamens and carpels. Both these reproductive organs are present in several whorls, and each organ is separate. The number of sepals is three or more, but the number of petals can range from none to several. For example, in prairie crocus (*Anemone patens*, Fig. 3.5), a member of this family, the sepals and petals are replaced by the large showy blue structures, the tepals, and the flower contains numerous stamens and carpels. Below the flower are leaf-like structures present in a whorl that form an involucre. The flowers of this family are hypogenous.

BRASSICACEAE

In the family Brassicaceae, which contains a large number of herbs and vegetables such as cabbage, mustard, radish, and the model plant *Arabidopsis*, flowers contain a fixed number of organs. In canola (*Brassica napus*), for example, there are four small sepals, inner to which are four large petals (Fig. 3.6). Stamens are present in two whorls; the outer whorl has two small stamens and the inner whorl has four long ones. The gynoecium consists of two carpels that are fused to form a pistil. The ovary is long with a short style and the stigma is flat, but in some cases it may be lobed. Brassica flowers are also hypogenous.

FIGURE 3.5 Flower of the prairie crocus (*Anemone patens*) showing the tepals, stamens, and carpels.

FIGURE 3.6 Canola (*Brassica napus*) flower showing sepals, four petals, and five stamens (one removed to show the pistil).

ROSACEAE

Flowers of the family Rosaceae are characterized by a cup-shaped hypanthium that may or may not be fused to the ovary. Sepals and petals are generally five in number, but the stamens are numerous and free from each other. The number of carpels varies, and they may be fused or separate. In the Saskatoon berry

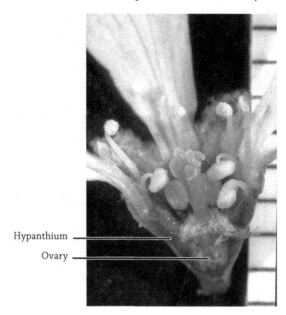

FIGURE 3.7 Flower of Saskatoon berry (*Amelanchier alnifolia*) showing cup-shaped hypanthium not fused with the ovary. (Courtesy of Dr. Richard St. Pierre)

(*Amelanchier alnifolia*, Fig. 3.7), sepals and petals are attached at the top of the hypanthium, stamens are in three whorls, and the ovary is syncarpous but the styles are separate, each with its own stigma. Flowers of this family are perigynous or epigynous.

POACEAE

Flowers of the grass family Poaceae are quite different from those described above. Flowers are generally small (called florets), they occur in groups of two or three in a spikelet, and they are enclosed in bract-like structures called glumes. Each floret is attached by an elongated stalk, the rachilla (Fig. 3.8). There are no tepals in grass flowers, but each floret has modified leaf-like structures, the lemma and the palea (see Fig. 3.8). The lemma is the larger of the two and often encloses the small palea. In many grasses, such as wheat, oat, and barley, the lemma has an elongated structure, the awn (see Fig. 3.8). There are three stamens in a floret, each with a filament and an anther, and at anthesis the filament elongates considerably, pushing the anthers out of the floret for pollen dispersal. The gynoecium consists of two carpels, the ovaries are fused, but the styles are separate and each has a feathery stigma. At the base of each floret, there are two or three swollen structures, the lodicules, which have a role

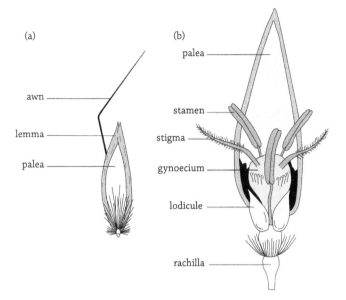

FIGURE 3.8 A stylized diagram of an intact (*a*) and dissected (*b*) grass floret showing various floral parts.

From *Botany: An Introduction to Plant Biology* by T. E. Weier, C. R. Stocking, and M. G. Barbour, 5th edition (1974), with permission of Dr. M. G. Barbour.

in the opening of the florets by separating the lemma and palea at the time of pollination.

THE INDUCTION OF FLOWERING

After a certain period of vegetative growth, variable in different species, and after receiving an appropriate signal—environmental (e.g., photoperiod or temperature) or chemical—the plant enters the flowering state. The induction of flowering involves major changes in the shoot apex, which stops producing leaves and converts into a floral or an inflorescence apex. Thus, a flower does not develop from a new growth center; instead, it is formed by the conversion of the vegetative shoot apex into a floral apex. Also, the vegetative apex loses its indeterminate nature and becomes determinate in terms of the fixed number of organs it produces. The flower may be terminal or axillary in position; in the latter it is the axillary meristem that forms the floral apex.

The transformation of the vegetative apex to a floral apex involves molecular, cytological, and structural changes in the shoot apical meristem (SAM). The early events involve enlargement of the SAM, which is achieved by both cell division and cell enlargement. The amount of increase varies, and in some cases there is a

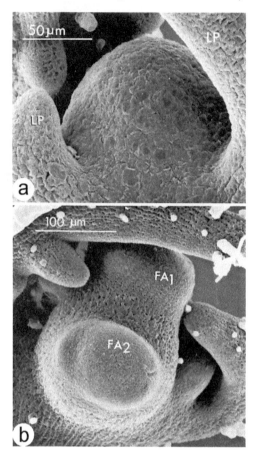

FIGURE 3.9 Vegetative (*a*) and flowering apices (*b*) of tomato showing the transformation from the dome-shaped to flattened apex. FA_1 and FA_2 = floral primordia 1 and 2, LP = leaf primordium.
From Sekhar and Sawhney (1984), *Canadian Journal of Botany* 44: 2404, with permission of the NRC Research Press.

substantial increase in size; for example, in the inflorescence apex of *Chrysanthemum* the apex enlarges as much as 300 to 400 times. The shape of the apex also changes during the transition to flowering. For example in tomato, the dome-shaped vegetative apex (Fig. 3.9a) begins to flatten as it enlarges (see Fig. 3.9b). In wheat and corn, the apex elongates considerably and becomes conical in shape. There are also several cytological changes in the cells of the floral apex, such as an increase in ribosomal density and an increase in the DNA and protein synthesis. In recent years there has been extensive research conducted into the genetic and molecular control of flowering. A number of genes have been identified that are expressed before and during the flowering process, and their regulation by environmental and hormonal factors has been investigated. The induction of flowering is a very active area of research in plant biology, and studies on a number of plant species have shown that genetic, molecular, hormonal, nutritional, and environmental factors all play important roles in controlling the flowering process. For further information on this topic, consult a plant physiology text.

FLOWER DEVELOPMENT

The development of a flower (i.e., the initiation and subsequent growth of different sets of floral organs) is also varied between species. The first set of organs, initiated at the margin of the enlarging floral apex, is the sepals. In tomato, for example, the first sepal primordium is formed as an outgrowth at the periphery of the apex and the subsequent sepal primordia are initiated in a helical fashion at an angle of approximately 137° from the previous one. The petal primordia are generally initiated after the sepal primordia and inner to them, and in most cases they are positioned alternate to the sepals. In tomato, petal primordia are initiated simultaneously in a whorl, but in other species their origin may be helical.

Stamen primordia are formed inner to petals and, in most cases examined, are alternate to petals, but opposite the sepals. In Brassicaceae, two whorls of stamens are formed and the primordia in the inner whorl are generally initiated before those in the outer whorl. In Rosaceae, several whorls of stamens are formed one after another. The last and the innermost organs to be formed are the carpels. In a syncarpous ovary, such as tomato, the carpel primordia are initiated around the remaining apex and grow over the apex and enclose it. The center of the apex grows upward and forms the placental tissue on which the ovules are initiated. Thus, the apex is ultimately used up in the formation of the flower. With the formation of each whorl of floral organs, the apex enlarges to make room for the initiation of the next set of organs. From this description it is evident that the initiation of floral organs is typically sequential and centripetal, from the outside to the inside. Although this pattern is found in many species, in some plants, such as canola, the pattern of floral organ initiation follows a slightly different sequence: sepals, stamens, petals, and carpels. This too emphasizes the diversity in angiosperm flowers.

The initiation of floral organs is followed by their growth. In general, sepals curve over the floral apex to protect the inner organs that are produced later. The later growth of sepals is upright, and at anthesis sepals bend outward for the opening of the flower. The petal primordia may or may not begin growth soon after initiation. In canola, for example, the growth of petals is delayed and occurs after that of stamens. Petals grow at a more or less constant rate during development, but in many cases there is a dramatic increase in petal length near anthesis (e.g., in *Petunia*), and this accompanies the opening of the flower. Similarly, in some plants (e.g., wheat), stamens also exhibit an accelerated rate of filament elongation just before anthesis, pushing the anther out of the floret to facilitate pollen dispersal. The carpel primordia generally grow at a steady rate throughout development.

In unisexual flowers, such as cucurbits and corn, growth and development of one of the two reproductive organs, stamens or carpels, is suppressed after initiation. In male flowers, the growth of carpel primordia is inhibited after initiation, whereas in female flowers stamen primordia are aborted soon after initiation. The pattern of initiation and growth of floral organs varies among species; for further information consult Greyson (1994) and Tucker (2003).

4

Reproduction

AS IN ALL organisms, reproduction is an essential aspect of the life of vascular plants, for only by this means is the continuity of a species ensured. In vascular plants the basic mechanism of reproduction is a sexual one in which two gametes unite to form a zygote, the first cell of the next generation. The process also includes meiosis or reduction division so that the repeated doubling of chromosome number is avoided. The linking of gamete fusion and meiosis in the life cycle also enhances the occurrence of genetic variation, which permits the operation of natural selection and hence evolution. The combining of different traits in gamete fusion and their random assortment in meiosis play an essential role in the evolutionary process. In Chapter 1, it was pointed out that in the vascular plants these two events take place in separate and regularly alternating generations. A diploid, asexual sporophyte produces haploid spores following meiosis and the spores give rise to haploid gametophytes, which produce the gametes, egg and sperm. Thus an asexual and a sexual generation alternate but the overall process is a sexual one. In the angiosperms this reproductive process occurs in the flower, and the structure of this modified shoot with some of its many variations was described in Chapter 3. It is now necessary to examine the actual reproductive process as it occurs in the reproductive organs. There is, however, another kind of reproduction that is widespread among the vascular plants. This is vegetative or asexual reproduction, and this alternative method will be examined briefly.

VEGETATIVE REPRODUCTION

The essence of vegetative reproduction is that a portion of the vegetative body of a plant directly gives rise to a new individual, which may remain attached to the parent for a time but ultimately becomes a separate entity. The ability to reproduce by way of the vegetative body is a reflection of the capacity of the plant for continued growth as well as regeneration—that is, the replacement of missing parts. In the broad context of reproduction this is a supplementary mechanism, but it is often very significant quantitatively and in some instances is the dominant means of propagation. Some plants that reproduce sexually in the major part of their geographic ranges increasingly rely upon vegetative propagation near the margins of those ranges where environmental conditions are less favorable. Vegetative reproduction is also of great practical significance, and many agricultural and horticultural varieties are propagated in this way.

In contrast to the sexual process, vegetative reproduction yields offspring that are genetically identical to the parent. Thus a particularly well-adapted individual may propagate itself rapidly without the variation that normally occurs in sexual reproduction. In agricultural practice this permits the propagation of useful varieties that are not genetically pure lines. The offspring of a single individual resulting from vegetative reproduction constitutes what is called a clone. A naturally occurring clone may become very extensive, covering many acres of area in some cases, for example the bracken fern (*Pteridium* sp.) in the United Kingdom or the poplar tree (*Populus* sp.) in North America. Similarly, a particular horticultural variety that is widely distributed and may have existed for centuries constitutes a clone. In spite of the immediate advantages associated with vegetative reproduction, there are long-term disadvantages in the absence of genetic variation that permit adaptation to new circumstances, in other words evolution. Not surprisingly, most species that reproduce asexually also reproduce sexually, at least sometimes.

Natural vegetative reproduction is accomplished by a variety of mechanisms. Plants that produce horizontal, trailing or vine-like shoots (e.g., squash) may form roots along the stem, often at the nodes, and such rooted shoots may become separated. Roots formed in this way (i.e., other than in the typical location) are called adventitious. Often the basal branches of woody plants may become buried and form adventitious roots in a process known as layering. Some plants with horizontal spreading roots may form adventitious shoot buds (e.g., the weed leafy spurge [*Euphorbia* sp.]), and later detachment results in new individuals. This process may be enhanced if roots are accidentally severed, and some pernicious perennial weeds are difficult to eradicate because of this capacity. Plants with

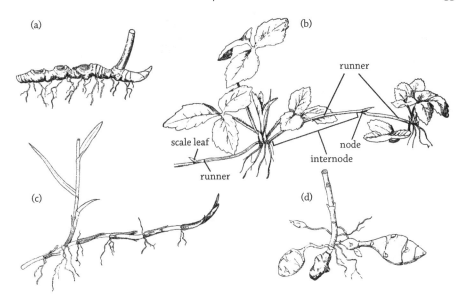

FIGURE 4.1 Different types of vegetative propagation: (*a*) rhizome (underground stem) of Solomon's Seal, (*b*) runner of strawberry, (*c*) stolon of witch grass, (*d*) tubers (swollen underground stems) of potato (*Helianthus tuberosum*).
From *Botany: An Introduction to Plant Biology* by T. E. Weier, C. R. Stocking, and M. G. Barbour, 5th edition (1974), with permission of Dr. M. G. Barbour.

horizontal underground stems or rhizomes (Fig. 4.1a) often spread extensively (e.g., the bracken fern or the *Canna* lily), and natural severance of the connection to the parent plant results in the new individual. Specialized horizontal lateral shoots, called runners if above ground (e.g., strawberry [see Fig. 4.1b]) and stolons if underground (see Fig. 4.1c), can also result in vigorous propagation. The formation of swollen underground stems called tubers (e.g., the potato [see Fig. 4.1d]), serve to perpetuate the plant but also result in propagation. Similarly enlarged roots of such plants as the sweet potato serve the same function. Bulbs and corms are swollen underground stems that can divide and produce additional shoots and form new individuals. Both bulbs and corms have stored food, bulbs in the leaves (e.g., the onion or garlic) and corms in the stem (e.g., *Gladiolus*). The leaves of some plants, notably members of the stonecrop family (Crassulaceae), give rise directly to small plantlets that either after detachment or while still attached to the parent plant form a new plant (e.g., Mexican hat [*Kalanchoe* sp.]). The structure of some of these specialized organs will be considered in later chapters.

Artificial propagation of economically important plants makes use of all of these natural mechanisms and often includes steps to enhance their occurrence, as in the artificial layering of woody plants by burying lower branches. The artificial rooting of pieces of the shoot called cuttings is a widely practiced technique.

Similarly, root cuttings may be used, as are detached leaves or pieces of leaves. Perennial plants that tend to form clumps, either by the development of basal branches or by adventitious buds on roots, may be divided artificially to produce new individuals. In more recent times, the use of sterile tissue culture on nutrient medium has become a widespread technique for propagating horticultural plants and trees, including fruit trees. Shoot apices explanted under sterile conditions often grow vigorously in culture, forming adventitious roots and numerous lateral shoots that may be separated and grown into mature plants. In this way prodigious numbers of new individuals can be produced quickly and in a limited space. Similarly, cell suspension cultures have been used to induce a large numbers of somatic embryos capable of developing into new plants, and this approach has also been used extensively in plant propagation. Some aspects of somatic embryogenesis will be considered in Chapter 5.

SEXUAL REPRODUCTION

The life cycle consisting of two alternating generations that is characteristic of all vascular plants was illustrated in Chapter 1 by a typical fern in which the two generations are independent and free-living organisms. The angiosperm plant with its flowers is a diploid sporophyte, and the occurrence of meiosis in the stamens and carpels results initially in the formation of spores. There are, however, significant differences from the fern reproductive system.

In most ferns all of the spores are of the same size and type (homosporous), and, upon release, each is capable of germinating to form a gametophyte that produces both eggs and sperms. In some other less advanced vascular plants the spores are of two types, and the plants are said to be heterosporous. In a cone of *Selaginella*, the little club moss, there are sporangia of two kinds: megasporangia (which contain a few, usually four, large spores or megaspores) and microsporangia (which produce a large number of small microspores). Before the spores are released, gametophytes begin to develop inside them, and the process is completed after shedding. The microspores develop into male gametophytes, which produce and release only sperms. The larger megaspores develop female gametophytes, which produce eggs in archegonia. Following fertilization the embryo (sporophyte) develops in the archegonium and is supplied by the nutrients stored within the megaspore and incorporated into the female gametophyte. The significance of the size difference associated with the sexual differentiation of the gametophytes and their retention within the spores is evident at this point. Efficiency is achieved by placing a larger reserve of nutrients in the megaspore for the female gametophyte, which must nurture the embryo, and a much smaller supply in the microspore for the male gametophyte, which need only produce and release the sperm. There is still a requirement for water, however, for the sperm to be transferred to the archegonium.

The reproductive system of the seed plants, known as the seed habit, accomplishes the transfer of the male gamete to the female without the necessity for a watery medium as in the case of the less advanced plants such as ferns. The megaspore is not released but is retained within the megasporangium or nucellus, which, protected by one or two covering layers or integuments, constitutes the ovule. Within the ovule the female gametophyte develops and the egg is formed. The microspores are produced in the anther of a stamen and develop into male gametophytes or pollen grains and are released from the anther. Pollen grains are transferred in the process of pollination to the carpel, and the sperms enclosed in the pollen are transferred to the egg in the ovule internally via the formation of a pollen tube. Thus the seed habit represents a highly modified heterosporous alternation of generations with a high degree of independence from the environment.

Development and Structure of the Ovule and the Female Gametophyte

In the angiosperms the ovules are formed within an ovary, a closed structure that is the basal swollen part of the carpel or pistil. The ovule begins with the outgrowth of the nucellus from the ovary wall. The nucellus has a short stalk or funiculus, and from this one or two layers of tissue, the integuments, are formed around the nucellus, enclosing it except for a tiny opening at the tip, the micropyle. This integumented nucellus is the ovule. It may remain erect, but more often it becomes inverted so that the micropyle is directed toward the point of attachment and the stalk is fused to the integument on one side.

The events that occur within the nucellus or megasporangium leading to megaspore formation and the development of the female gametophyte vary, but one pattern has been found to be the most common, the *Polygonum* type. In this case, one cell within the nucellus enlarges and functions as the megaspore mother cell (Fig. 4.2a). This cell undergoes meiosis, resulting in a tetrad of haploid megaspores (see Fig. 4.2b) that have a linear arrangement parallel to the axis of the ovule. The three megaspores nearest the micropyle degenerate, leaving the one at the opposite or chalazal end of the ovule as the functional megaspore (see Fig. 4.2c). This megaspore enlarges to form the female gametophyte, also known as the embryo sac, and three successive mitoses result in the formation of first two, then four (see Fig. 4.2d), and finally eight nuclei. Four of these nuclei are located near the

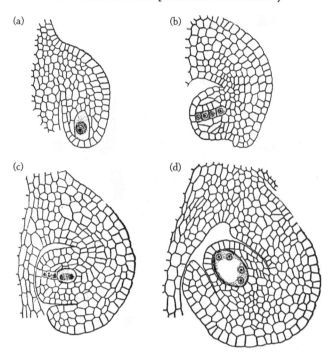

FIGURE 4.2 Development of the female gametophyte (embryo sac) in *Anemone patens*: (a) megaspore mother cell, (b) four haploid megaspores and developing integument, (c) one functional megaspore at the chalazal end, (d) four haploid nuclei within the developing embryo sac.

micropylar end and four at the chalazal end. At the micropylar end three of the nuclei are segregated in discrete cells within the embryo sac. These constitute the egg apparatus consisting of the egg and two synergids (Fig. 4.3). At the opposite end three nuclei are also segregated into discrete cells called the antipodals. The two remaining free nuclei, one from each end of the embryo sac, migrate toward the egg apparatus and are known as the polar nuclei. In some plants the polar nuclei fuse to form the diploid fusion nucleus. The remainder of the embryo sac not included in the egg apparatus and the antipodals and containing the polar nuclei or the fusion nucleus constitutes the central cell (see Fig. 4.3).

The embryo sac is thus a much reduced female gametophyte. The three cells of the egg apparatus, the egg and the two synergids, are placed in a triangular arrangement as seen in cross-section, and the synergids extend to the end of the embryo sac while the egg is slightly displaced toward the interior (i.e., away from the micropyle). The cells of the egg apparatus are covered by a thin cell wall that usually fades out or becomes honeycombed toward the interior. At the micropylar end, each synergid has fingerlike ingrowths of the cell wall known as the filiform apparatus (see Fig. 4.6a later in the chapter). This consists of a thickened

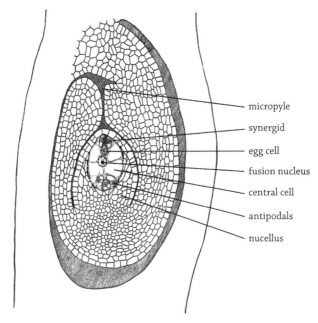

FIGURE 4.3 Fully developed embryo sac ready for fertilization. At the micropylar end an egg and two synergids are present, three antipodals at opposite (chalazal) end of embryo sac, and the two polar nuclei in the central cell have fused to form a diploid fusion nucleus.

mass of soft wall material that in effect greatly increases the surface area of the cell membrane and is believed to facilitate the passage of molecules into or out of the synergids. It also plays an important role in the fertilization process, as discussed later. The antipodals at the opposite end of the embryo sac also have full or partial walls and may have delicate, inwardly projecting wall projections that increase the surface area of the cell membrane, thus enhancing the passage of molecules, most probably nutrients entering. In some cases the antipodals disappear quickly; in others they increase in number to as many as 100 in barley, for example. The entire embryo sac is surrounded by a cell wall that is effectively the wall of the central cell, and this wall also may have inward wall projections, particularly near the egg apparatus. At the micropylar end this wall is fused with those of the egg and synergid cells, and there is no wall on the central cell side where these cells extend into it.

Although the pattern of female gametophyte formation and structure described above is the most common, at least 12 other types have been described. The types are ordinarily named for the genus in which they were first described. Thus the common type described above is known as the *Polygonum* type. In the *Allium* (onion) type, after the first meiotic division of the megaspore mother cell, one of the two resulting cells degenerates and the remaining cell undergoes the second division to form two haploid nuclei, in effect the nuclei of two megaspores. Now these nuclei divide twice, and the cell with eight nuclei is organized into an embryo sac as in the *Polygonum* type. Because the female gametophyte develops from the equivalent of two megaspores it is called bisporic. An even more striking departure is found in the *Fritillaria* type, which includes *Lilium*, most often used to demonstrate angiosperm reproduction to students. *Lilium* is tetrasporic because meiosis in the megaspore mother cell results in four haploid megaspore nuclei without cytokinesis (wall formation), and these four nuclei go on to organize the embryo sac directly. However, the way in which this is accomplished is rather interesting. One nucleus is located near the micropylar end of the developing embryo sac and three are placed at the opposite end. In the next division the solitary nucleus forms two haploid nuclei, but the three merge as they divide and give rise to two triploid (three chromosome sets) nuclei. Each pair of nuclei now divides again. The four haploid nuclei at the micropylar end form the egg, two synergids and one haploid polar nucleus. The four triploid nuclei at the opposite end form three antipodals and one triploid polar nucleus. When the polar nuclei ultimately fuse, the resulting nucleus will be tetraploid (four chromosome sets). The remaining embryo sac types examined are mostly tetrasporic and vary mainly in the number of cells and nuclei formed and their positions within the embryo sac. For example in *Plumbago*, no synergids are developed and the egg stands alone at the micropylar end of the embryo sac. In others, the number of polar nuclei is increased to four or eight. In all cases, however, the basic reproductive cycle is not disrupted in that the egg that transmits the parental genes to the next generation is derived from one megaspore. For further description of these patterns please consult Maheshwari (1950) and Yadegari and Drews (2004).

Development and Structure of Pollen Grain or the Male Gametophyte

As described in the previous chapter, the anther of a stamen commonly includes four pollen sacs, although fewer or more are found in some species. A pollen sac is a microsporangium that contains numerous microspore mother cells (MMCs). Each MMC divides by meiosis to form a tetrad of haploid microspores (Fig. 4.4). The tetrads are initially held together by callose (β-1,3 glucan) wall, but this wall is later dissolved by the enzyme callase, which is secreted by the tapetum, the innermost layer of the anther. At this stage the microspores have a few small vacuoles, but then they become highly vacuolated, each with a large vacuole. At the same time, microspores develop an additional thick wall called the exine, which is outside the inner cell wall, the intine. The microspores divide by mitosis,

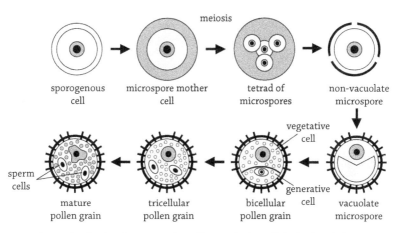

FIGURE 4.4 Stages in the development of a pollen grain (tricellular) of *Arabidopsis thaliana*. The mature pollen grain has a vegetative cell and two sperm cells.
From Fei and Sawhney (2001), *Canadian Journal of Botany* 79: 118–129, with permission of NRC Research Press.

but the division is unequal and a curved thin wall separates a large vegetative cell from a small generative cell, and both these cells now have many small vacuoles (see Fig. 4.4). This is the bicellular pollen grain or the male gametophyte, and in some plants (e.g., tomato, a member of Solanaceae), it is released from the anther at this stage. In other species (e.g., *Arabidopsis*, a member of Brassicaceae, or the grass family Poaceae), the generative cell divides to form two sperm cells and the pollen grain is tricellular at maturity (Figs. 4.4 and 4.5)—that is, at the time of dispersal from the anther.

The pollen wall is a complex structure consisting of different layers and has various wall components. The exine consists of an outer part, the sexine, which may be smooth or have rod-like structures, the bacula, and the inner part, the nexine, which may be of one or two layers. The exine is a tough wall made of cellulose, and it contains a complex polymer, sporopollenin, which is extremely resistant to decay and helps both in the preservation of pollen grains and in protection from invasion of pathogens. Thus, as pollen grains are very small (i.e., two- or three-celled) and are free-floating independent structures, the exine plays a very important role in their survival and in turn of the sperm cells contained in them. The exine also contains a number of proteins that have roles in the compatibility response as well as in pollen germination. The exine is also often beautifully sculptured with various patterns that are characteristic of a species. By contrast, the intine is relatively simple, similar in structure to the primary cell wall, and it does not contain sporopollenin. The pollen wall has thin areas, called the pores, through which the pollen tube will emerge at the time of

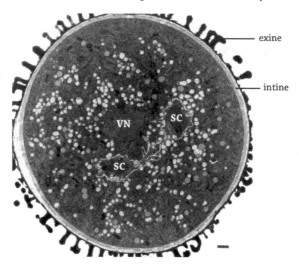

FIGURE 4.5 Mature tricellular pollen grain of *Arabidopsis thaliana* with a vegetative cell and its nucleus (VN), two sperm cells (SC), and an outer exine and inner intine wall.
From Fei and Sawhney (2001), *Canadian Journal of Botany* 79: 118–129, with permission of NRC Research Press.

germination. For further information on pollen development and characteristics of mature pollen see Bedinger (1992) and Shivanna (2003).

Pollination and Fertilization

After their release from an anther, pollen grains are carried by insects, birds, or wind to the stigma of a carpel. The transfer of pollen from one plant to the flower of another is called cross-pollination, and to the carpel of the same flower is self-pollination. Cross-pollination leads to genetic diversity in a population, which is important for the evolutionary process. In fact, some plants are self-incompatible—that is, pollen grain either does not germinate on the stigma of the same flower or if it does germinate, the pollen tube growth is stunted. Thus, self-incompatibility ensures cross-pollination and, therefore, has an important role in agriculture for the production of hybrid crops.

After pollen grains land on the stigma of a carpel, they germinate and put out a pollen tube, which grows out through a pore in the wall; the intine forms the initial cell wall of the pollen tube. Pollen germination requires a high sugar level, and the stigmatic cells exude a number of substances, including sugars. There is also a complex recognition mechanism involving proteins in the pollen wall and the stigmatic surface, and this serves to eliminate stray pollen of other species and, in many cases, to enhance cross-fertilization by preventing the germination of grains of the same genetic constitution. These proteins reflect the genetic makeup of the respective parent plants and, unless they interact in the correct fashion, germination is prevented or at least retarded.

After germination, the pollen tube, growing at its tip, penetrates the stigma and extends through the style toward the ovules in the ovary. The style may contain an open canal filled with mucilaginous substances through which the tube grows. Alternatively, the style may have a core of specialized cells called the transmitting tissue, and the tube grows by either penetrating between the cells or growing through the soft cell walls. The pollen tube grows rapidly in some cases, reaching the ovule within minutes or hours, but in other species weeks or months may be required. Depending upon the length of the style, the extent of tube growth may be limited or substantial, for example up to 50 cm in length in the silk (i.e., the style) of corn ear. As the pollen grain germinates, the vegetative nucleus and the sperm cells move into the tube and remain localized near the growing tip. If the generative cell has not divided before shedding from the anther, it divides and produces two sperm cells during pollen germination or tube growth.

To effect fertilization, the sperm must be delivered to the egg in the embryo sac, and in this process synergids play an important role. Upon reaching the ovule, the pollen tube grows through the micropyle of the ovule and penetrates the nucellus, which may partially break down in that region. At the time of penetration of the pollen tube, or in some cases before the entry of the pollen tube, one of the synergids begins to degenerate (i.e., its contents lose their organization) and the pollen tube grows into the degenerating synergid. The pollen tube enters through the filiform apparatus, the soft wall material, of the synergid (Fig. 4.6a). Within the degenerating synergid the pollen tube grows briefly. It then stops growth, is ruptured near the tip, and then discharges its contents, including the two sperm cells (see Fig. 4.6b). One of the sperms fuses with the egg (see Fig. 4.6c), and its nucleus merges with the egg nucleus to form the diploid zygote. The zygote is the first cell of the new sporophyte generation. The second sperm joins with the central cell and its nucleus fuses with the two polar nuclei, or more often with the fusion nucleus if they have already joined, to form the triploid primary endosperm nucleus (PEN). Thus, both the egg and the central cell (with polar nuclei) are fertilized. This process is called double fertilization and is essentially unique to the angiosperms except for a similar process leading to additional embryos in the gymnosperm group Gnetales.

The PEN initially divides by nuclear divisions and then cell wall formation occurs to form a cellular endosperm, the highly nutritive tissue that contains, sugars, amino acids, proteins, and various minerals and hormones; these

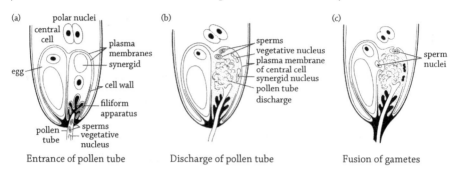

FIGURE 4.6 Stages in the fertilization process in an angiosperm embryo sac: (*a*) entrance of pollen tube, (*b*) discharge of pollen tube contents, (*c*) fusion of male and female gametes. From "The Embryo Sac and Fertilization in Angiosperms" by W. A. Jensen, lecture no. 3 (1972), with permission of Harold Lyon Arboretum Publication.

substances are later used by the developing embryo. The endosperm may persist in the mature seed, as in grasses, and it serves as a source of nutrients for the germinating seed. Indeed, it may occupy as much as two-thirds of the seed in some grasses (e.g., wheat and corn). In many dicots, the endosperm is used up during seed maturation, as in legumes and tomato, and the mature seed is devoid of endosperm. In species that have a *Polygonum* type of embryo sac the endosperm is triploid, but in some other types it may be of a higher ploidy. For example, in the *Fritillaria* type (described above) where one polar nucleus is haploid and the other is triploid, after fusion with the sperm the endosperm becomes pentaploid (five sets of chromosomes).

OVERVIEW OF SEXUAL REPRODUCTION

In essence, while the angiosperms maintain the basic pattern of alternation of generations in their life cycle, they show significant departures from the plan illustrated in the ferns (as described in Chapter 1). Spores are produced, but they do not function in dispersal. Rather, this mechanism has been adapted and microspores, which develop into pollen grains, serve as a means of transferring the male gametes to the female reproductive organ. The enclosure of the ovules in the ovary not only protects these delicate structures but also permits the operation of a screening mechanism that selects the male gametophytes that are permitted to deliver male gametes. The gametophytes are much reduced in size and cell number and are highly specialized in the performance of a complex fertilization process. Further, the introduction of a second fertilization interjects a novel nutritional tissue, the endosperm, into the basic life cycle.

In the seed plants other than angiosperms, collectively the gymnosperms, some of these specializations are lacking. Here the ovules are not enclosed (are naked) at the time of pollination and the pollen is transferred directly to the ovules; it is drawn into the micropyle and germinates inside the ovule. The female gametophyte undergoes substantial development, becomes filled with nutrients, and nurtures the embryo as it develops. It produces eggs in vestigial archegonia. Two sperms are produced by each male gametophyte but there is no second fertilization; the role of the endosperm is performed by the female gametophyte. If, as is often the case, several pollen grains germinate within the ovule, more than one egg may be fertilized, but ordinarily only one embryo survives to germinate. Nevertheless, the basic advantages of the seed habit are achieved.

APOMIXIS

There is another kind of reproductive mechanism that occurs in some angiosperms in which the production of seeds does not involve a sexual process. This phenomenon, known as apomixis, combines the essential characteristics of vegetative reproduction with the advantages of dispersal by seeds. In some cases the embryo sac is formed without the occurrence of meiosis; therefore, all of its cells are diploid. The egg, which is diploid, or another cell of the embryo sac forms an embryo without fertilization, as in dandelion (*Taraxacum*). In other instances (e.g., some *Citrus* species), the embryo develops from a cell outside the embryo sac, in the nucellus or integuments, and requires no fertilization because it is already diploid. Often pollination is necessary to stimulate embryo development even though no fertilization is involved, or it may be required for the other fertilization process, which leads to endosperm development. Apomixis without pollination, however, is also common. As in the case of typical vegetative reproduction, many apomictic species also reproduce sexually at least occasionally so that some genetic recombination does occur.

5

Embryo, Seed, and Fruit Development

AS INDICATED IN the preceding chapter, the zygote or fertilized egg marks the beginning of the sporophyte in the life cycle of a plant and develops into an embryo. The term *embryo* refers to a young sporophyte that is produced from the zygote and is dependent on the parent plant for nutrition. With the exception of algae, embryonic stages of the sporophyte are observed in all plants, including bryophytes (liverworts and mosses), vascular cryptogams (ferns and other less advanced vascular plants), gymnosperms, and angiosperms. In bryophytes and lower vascular plants, however, the embryonic stages are not sharply defined and the embryo continues uninterrupted growth to develop into a mature sporophyte. In seed plants, with few exceptions, the early stages of sporophyte development are clearly separated from the later stages by the formation of a dormant embryo. In these plants the embryo is encased in a seed coat, and the development of the seed takes place in concert with embryo development.

PATTERNS OF EMBRYO DEVELOPMENT

In angiosperms the pattern of embryo development is varied among different families. In addition, there are major differences between the structure and development of dicotyledon and monocotyledon embryos. Because of space limitation only one example from each of these two groups of flowering plants will be described in detail, but some indication of variations will be mentioned.

Dicotyledons

Although there are variations among dicotyledon species, the pattern of embryo development is reasonably consistent within a species and in many cases within a family. One of the species in which the structure and development of the embryo has been studied in considerable detail is *Capsella bursa-pastoris* (shepherd's purse), a member of the family Brassicaceae. In the model plant *Arabidopsis thaliana*, also a member of Brassicaceae, and in which there has been considerable research in the genetic and molecular control of embryo development, the pattern of embryo development is similar to that of *Capsella*. Here we will discuss mainly the development in *Capsella* embryos and describe some variations in other species.

In most plants, the first division of the zygote does not occur immediately after the fusion of male and female gametes. Commonly there is a gap of a few hours, and in some cases, as in cotton, 3 to 5 days elapse before the onset of cell division. The first division of the zygote (Fig. 5.1a) in almost all cases studied, including *Capsella* and *Arabidopsis*, is transverse. It is an unequal division resulting in a small terminal cell facing the chalazal end of the gametophyte and a large basal cell at the micropylar end (see Fig. 5.1b). This first division is the first step in embryo differentiation separating the terminal cell, which will form

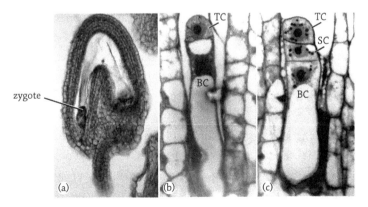

FIGURE 5.1 Embryo development in *Capsella bursa-pastoris*: (a) zygote enclosed in an ovule, (b) two-celled embryonic stage, (c) three-celled embryo. BC = basal cell, SC = suspensor cell, TC = terminal cell.

Figures b and c from Schulz and Jensen (1968), *American Journal of Botany* 55: 807–819, with permission of the Botanical Society of America).

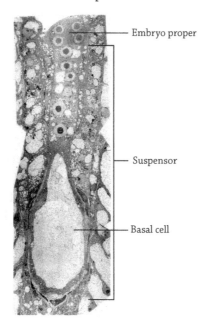

FIGURE 5.2 Young embryo of *Capsella bursa-pastoris* attached to a suspensor with a large basal cell.
From Schulz and Jensen (1969), *Protoplasma* 67: 139–163, with permission of Springer Science and Business Media.

the embryo proper, and the basal cell, which mainly forms the suspensor. The terminal cell in *Capsella* is cytoplasmically dense with a few small vacuoles, in comparison to the basal cell, which possesses a large vacuole and a thin layer of cytoplasm. The next few divisions are in the basal cell and are transverse, first forming a suspensor cell (see Fig. 5.1c) and then producing a filamentous suspensor with six to eight cells (Fig. 5.2). The basal cell of the suspensor is considerably enlarged with a large vacuole and is attached to the embryo sac wall at the micropylar end. The uppermost cell of the suspensor (i.e., next to the terminal cell) is the hypophysis cell, which also contributes to the development of the embryo, including the formation of the root cap, and cortex and epidermis of the root.

After the suspensor is formed, the terminal cell divides, and the first division of the terminal cell is vertical and results in a two-celled embryo, the diad (Fig. 5.3a). The second division is also vertical, resulting in a four-celled embryo, the quartet (see Fig. 5.3b). The four cells then divide transversely, forming an eight-celled (octant) embryo (see Fig. 5.3c [only four cells are visible]). It is important to note that although the cells of the octant embryo appear structurally and cytologically similar, some biochemical and molecular differences separate the upper four cells from the lower cells. The eight cells of the octant

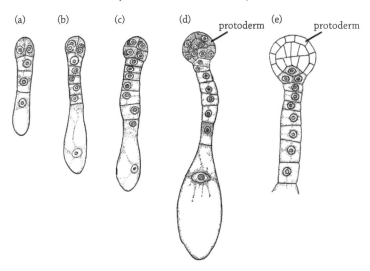

FIGURE 5.3 Stages in the development of embryo proper of *Capsella bursa-pastoris*: (*a*) diad, (*b*) quartet, (*c*) octant, (*d* and *e*) differentiation of the protoderm and formation of globular embryo.
From *Plant Anatomy*, 4th edition (1990), by A. Fahn, with permission of Elsevier Ltd.

next divide periclinally (i.e., parallel to the surface) to form a surface layer of cells. This surface layer is the first stage in the formation of the protoderm, the precursor of the epidermis, and is the first visible sign of differentiation in the embryo (see Fig. 5.3d). The inner cells of the octant embryo divide longitudinally, resulting in upper and lower tiers of 16 cells each. Cell divisions also take place in the protoderm, but they are anticlinal (i.e., perpendicular to the surface). All these divisions result in a ball of cells called the globular embryo or proembryo (see Figs. 5.3e, see also Fig. 5.5a). The inner cells of the globular embryo further divide longitudinally, forming long narrow cells that constitute the procambium cells (Fig. 5.4), which will give rise to the future vascular tissues, the xylem and phloem. The procambial cells are surrounded by larger cells that are more vacuolated than the procambial cells, and these cells form the ground meristem (see Fig. 5.4). The ground meristem is the precursor of the ground tissue or the fundamental tissue.

Simultaneous to the early stages of cell differentiation, the embryo changes shape from radial to bilateral symmetry—in other words, it flattens. Further cell divisions are localized at two regions at the chalazal end of the embryo, resulting in two outgrowths that will form the future cotyledons. Because of these localized growths the embryo now assumes a heart shape (Fig. 5.5b). At the heart-shaped stage the embryo begins to turn green and becomes photosynthetic. The embryo body next elongates, forming a short axis, the hypocotyl, and at the

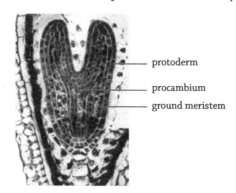

FIGURE 5.4 Torpedo stage of an embryo in *Capsella bursa-pastoris* with the early differentiation of the protoderm, procambium, and ground meristem.

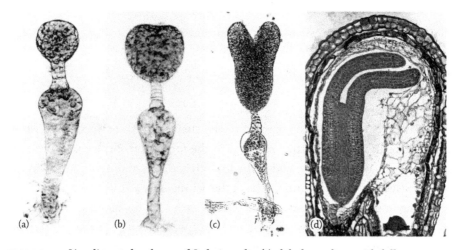

FIGURE 5.5 Live dissected embryos of *Stelaria media*: (a) globular embryo with fully developed suspensor, (b) heart-shaped embryo, (c) torpedo stage, (d) a section of *Capsella bursa-pastoris* ovule with "walking-stick" stage embryo.

same time there is further growth of cotyledons, resulting in the torpedo-stage embryo (see Figs. 5.4 and 5.5c). Subsequent growth of cotyledons and the embryonic axis results in bending of cotyledons, because of the restricted space in the ovule, and the embryo is now at the "walking-stick" stage (see Fig. 5.5d). During further growth, the shoot and root apical meristems (SAM and RAM) become established in the embryo; the SAM is located between the two cotyledons and the RAM at the opposite end of the embryo (Fig. 5.6). It should be noted that the cotyledons are not produced by the SAM and, therefore, are not truly leaf-like in origin. The mature embryo fills the ovule (now a seed), having absorbed most of the storage material of the endosperm. Further growth of the embryo and the

Embryo, Seed, and Fruit Development

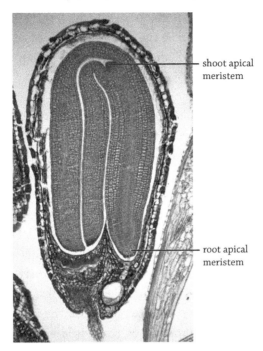

FIGURE 5.6 Mature embryo of *Capsella bursa-pastoris* with shoot and root apical meristems, cotyledons, and a root-shoot axis. (Courtesy of Dr. Ed Yeung)

seed is arrested and the seed now enters a state of dormancy (i.e., inactivity). The mature seed is a highly desiccated (less than 10% water) structure and the embryo can remain in the dormant state for several months, indeed in some species for several years.

Although *Capsella* illustrates the general pattern of embryo development in the dicotyledons, there are differences within this large group. These differences include variations in cell division patterns after the first division of the zygote, the origin of the suspensor and embryo proper, and the extent of suspensor development. These differences, however, do not obscure the significance of the establishment of SAM and RAM in the embryo, which will later form the shoot and root systems of the primary plant body.

Monocotyledons

The major difference in the monocotyledon embryo from that of dicotyledons is that it has a single cotyledon (*mono* = single). The early development of a monocot embryo is, in many ways, similar to that of dicots, but at the stage of cotyledon initiation differences begin to appear; there is the formation of a single cotyledon in an apparently terminal position. The SAM is located in a lateral position, but this is believed to be the result of displacement by the enlarging single cotyledon. Embryo development in maize (*Zea mays*) will be used as an example.

In maize the suspensor is multicellular and is more than one cell wide. Five days after fertilization the embryo becomes club-shaped—that is, the distal part is enlarged, which is the embryo proper, and the basal part is the suspensor. On one side of the embryo there is greater growth than on the other side, and there is formation of a single cotyledon, also called the scutellum (shield). The growth of the scutellum is extensive; it grows upward, toward the base, and around the epicotyl axis located on the side opposite to the scutellum (Fig. 5.7). The epicotyl axis is enclosed in a thin layer of tissue called the coleoptile (see Fig. 5.7). The coleoptile surrounds the epicotyl and protects it during its subsequent development. As the epicotyl grows it changes its position from lateral to vertical. In the epicotyl the shoot apex is differentiated, including the SAM, which produces two or three leaf primordia. At the lower end of the embryo axis, the RAM with a root cap is formed. The root apex and the young root, the radicle, are also enclosed in a thin layer of tissue called the coleorhiza (see Fig. 5.7). In grasses the mature embryo also contains the first internode, which is located between the scutellum and the coleoptile and is called the mesocotyl. In some cases a root primordium is differentiated from the internode (adventitious root) and is present in the mature grass embryo.

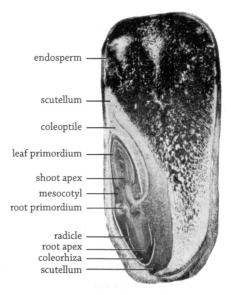

FIGURE 5.7 A mature kernel of maize (*Zea mays*) with embryo and endosperm.

It would be evident that the structure of mature embryo in corn is quite different from that of *Capsella* described earlier. Another difference is that whereas in *Capsella* the endosperm is used up during embryo development, in maize the endosperm persists at maturity. Indeed in grasses, including most cereals, the endosperm occupies as much as two-thirds of the seed (see Fig. 5.7).

The Role of the Suspensor

The suspensor is varied in structure in different species, but it is an important structure for embryo development. The name *suspensor* was given because it "suspends" the embryo in the female gametophyte and the ovule while attached to the parental tissue. There are several roles assigned to the suspensor, including that of anchorage, and the absorption of nutrients from the surrounding tissues (the endosperm and the nucellus) and their transport to the developing embryo. Another suggested role of the filamentous suspensor is to push the embryo deeper into the female gametophyte. However, this role is more applicable in gymnosperms than in angiosperms because in gymnosperms there is generally a large female gametophyte that provides nutrients to the embryo.

SOMATIC EMBRYOGENESIS

In Chapter 4, the subject of apomixis, the production of seeds with embryos but without the involvement of the sexual process, was described. With advancements in the techniques of tissue culture, the process of producing embryos from somatic (body) cells in culture (in vitro) has greatly increased. In most cases the cultures are of cell suspensions, or of protoplasts (without the cell wall), but the separation of the cells is not always necessary. Although the initial stages of embryo development in such cases are usually not identical to those in an ovule (e.g., the typical suspensor may be lacking), the embryos produced are similar in structure to zygotic embryos developed in seeds and can generally be grown to form mature plants. This phenomenon has been very important in the area of plant propagation, but more importantly it documents that differentiated plant cells do not lose their developmental potential and can be induced to form embryos.

Experimental studies have shown that embryos can also be produced from microspores and megaspores (haploid structures), and these are called haploid embryos. These embryos can develop into mature plants with flowers and functional gametes after diploidization of chromosomes by chemicals (e.g., colchicine) and are also called double haploids. The result is plants that have the same set of genes and chromosomes (homozygous) and are genetically uniform and, therefore, are very useful in agriculture. The ability of microspores to form embryos shows that even with one set of chromosomes certain plant cells have the ability to differentiate and develop into embryos.

SEED DEVELOPMENT
Endosperm Formation

As indicated in Chapter 4, double fertilization in angiosperms involves fusion of both the sperm cells in a pollen tube, one with the egg cell to form the zygote and the other with two polar nuclei to initiate endosperm development. The result of the latter fusion is the formation of a triploid primary endosperm nucleus (PEN). However, as discussed earlier, in some species (e.g., *Lilium*), the PEN is pentaploid. The PEN divides and initiates the formation of endosperm, the nutritive tissue that supplies many of the essential metabolites for embryo development. The pattern of endosperm development is also varied in different angiosperm species.

In *Capsella*, the PEN divides to form two nuclei that migrate to opposite ends of the central cell, one at the micropylar end and the other at the chalazal end. Both these nuclei then divide repeatedly, resulting in a multinucleate condition. The central cell is highly vacuolate and the nuclei become distributed in a thin layer of cytoplasm around its periphery. At the globular embryo stage, the endosperm nuclei aggregate to form nodules near the cell wall of the central cell. When the embryo reaches the late heart stage, cell wall formation begins to take place in the endosperm. However, soon after, cells of the endosperm start to break down and their contents are absorbed by the suspensor and transported to the developing embryo. In *Capsella* the endosperm is mostly used up during embryo development and in the mature seed only remnants of it are visible; the reserves are now contained within the embryo itself. This pattern of storage is particularly evident in the legume family (Fabaceae) in which the reserve nutrients are stored in the swollen cotyledons. The pattern of endosperm formation in *Capsella* is not the same in all dicotyledons and certainly not in most monocotyledons.

In another pattern of endosperm formation known as cellular, cell wall formation follows the first nuclear division of PEN and continues throughout endosperm development. In grasses (e.g., wheat and corn), there is extensive development of the cellular endosperm, which persists at maturity. In other cases (e.g., *Impatiens*), the cellular endosperm develops outgrowths called haustoria that invade the surrounding tissues. There is also a third pattern in which the embryo sac is divided unequally by the first division of the PEN: the derivative at the chalazal end develops without further cell wall formation while the smaller portion at the micropylar end may or may not become cellular, as in *Trillium* and *Juncus* species.

Seed Maturation

The ovule has one or, more often, two integuments surrounding the nucellus and the developing embryo within the ovary. After fertilization the integuments undergo changes to form the seed coat or testa, which serves as a protective covering for the developing embryo inside as the seed prepares for its dispersal to the external environment. In cases where a one-seeded fruit is shed as a unit, as in the case of achenes of the aster family or the caryopses of the grasses (Table 5.1), there is minimal integument development and the protection is provided by the differentiated wall of the ovary. However, where the seeds are shed individually or are contained inside a fruit from which they are ultimately released, seed coat development is more extensive. There may be several protective layers, including a layer of thick-walled sclerenchyma that constitutes the protective testa of the matured seed. In addition to its role of protection, the seed coat also serves an important function of seed dispersal. The seed coat may develop devices that favor dispersal, such as hairs on dandelion or cotton seeds, wings that favor wind dispersal (e.g., in maple), or flotation devices for water distribution. In some cases, the seed coat may develop spines or hooks that become attached to an animal's fur or feet, which carry the seed to long distances. Seeds consumed by animals are often passed out intact at the other end as the seed coat is resistant to the digestive enzymes of the animal, thereby helping in seed dispersal. Finally, mature seeds are rich in proteins, carbohydrates, and lipids and are an important source of food and oil, especially unsaturated fatty acids, for human and animal consumption.

Embryo and Seed Dormancy

The maturation of the embryo is followed by a period of dormancy of the embryo and the seed in which it is located. In other words, the embryo reaches a certain stage of development and then its further growth is halted. This developmental arrest is caused by a number of factors, but the major one is desiccation (i.e., loss of water in the seed and in the embryo). The water content in the seed is reduced from approximately 70% to 10% in the dormant seed. Another factor known to be associated with seed dormancy is the plant hormone abscisic acid (ABA); the

TABLE 5.1

COMMON FRUIT TYPES

I. **Simple Fruits**—derived from a single pistil
   Dry Fruits
      Dehiscent
         From a single carpel
            Follicle—opens on one side (milkweed, peony, columbine)
            Legume—opens on two sides (characteristic of the pea or legume family)
         From several united carpels (two or more)
            Capsule—from a compound pistil, may open by slits on backs of carpels (iris, lily), where carpels meet (St. John's wort), by pores near top (poppy) or by removal of top (Brazil nut)
            Silique—a two-locular capsule. Two valves separate from a central partition (Brassicaceae family)
      Non-dehiscent fruits containing a single seed
         Caryopsis—pericarp fused to seed coat (grass family)
         Achene—pericarp not fused to seed coat but close fitting (buckwheat)
         Utricle—pericarp inflated, bladder-like (goose foot)
         Nut—pericarp very hard and bony (acorn)
         Samara—fruit winged (ash) and may be double-winged (maple)
   Fleshy Fruits
      Drupe—stony endocarp around one seed (peach)
      Pome—derived from inferior ovary; hypanthium distinct from leathery ovary wall (apple)
      Berry—whole pericarp more or less fleshy; true berry from superior ovary (tomato, grape); false berry from inferior ovary (blueberry)
         Hesperidium—special berry with leathery exocarp (orange)
         Pepo—special berry with hard rind (squash)
II. **Aggregate Fruits**—derived from several pistils of one flower (raspberry)
III. **Multiple Fruits**—derived from pistils of several flowers (pineapple, mulberry)
IV. **Aggregate-Accessory Fruits**—structures other than pistils involved (strawberry, fig)

---

amount increases during seed maturation and its concentration reaches its maximum in mature seed.

The seed containing the dormant embryo is dispersed from the fruit; in plants that produce a single seed, it has the fruit wall attached to it. The mature dormant embryo is in fact a plantlet, a miniature plant, with the SAM and RAM and cotyledons, the leaf-like structures that will provide nutrients during the early stages of seedling growth. As noted above, in grasses the embryo is more advanced in development than in dicots, as it has two or three true leaf primordia and in some cases the first internode may also be differentiated. After dispersal, when the seed reaches a suitable environment (i.e., in terms of moisture and temperature), the embryo in the seed resumes growth and the seed germinates and develops into a young seedling. There may, however, be limitations imposed by dormancy, which may be no more than a failure of hydration caused by an impermeable seed coat, which is reduced over time, or it may be the result of a physiological inhibition, which also may be diminished or eliminated in time. Seed dormancy has significance for the survival of a species in that it tends to prevent germination under unfavorable conditions such as in the autumn before the onset of winter. In some species, such as willows and poplars, seeds germinate as soon as they reach the soil if moisture is available. Many seeds have a dormancy that is lifted by an environmental exposure (e.g., cold, heat, light or heavy rainfall related to the natural habitat of the species). In some plants, such as the mangroves, there is no seed dormancy and the embryo in a seed continues to grow and forms a plantlet while still attached to the parent plant; this condition is called vivipary.

FRUIT DEVELOPMENT

A fruit is defined as a ripened or matured ovary, but in general terms it includes other structures adhering to the ovary and may combine more than one ovary. It is difficult to organize fruit types into a useful classification scheme, but the system presented in Table 5.1 includes most of the common types. One distinction that is not made in this system is that between fruits derived from superior and inferior ovaries (see Chapter 3). The fruit types are grouped as simple fruits, aggregate fruits, and multiple fruits. A simple fruit is derived from a single pistil, either one carpel or several fused carpels. It may be either dry or fleshy, and if dry either dehiscent or non-dehiscent. The aggregate fruits are those in which separate carpels from the same flower form the fruit, as in the raspberry. In cases such as the strawberry, in which the receptacle bearing several carpels is included, the term *aggregate-accessory* may be used. In a multiple fruit such as the pineapple or the mulberry, the pistils of several flowers are included in a single fruit.

The wall of the fruit, known as the pericarp, represents the matured or ripened structure of the ovary. In many cases it represents the structure of a carpel or a compound pistil, but in the case of an inferior ovary there is believed to be non-carpellary tissue included in it. There is considerable discussion about the nature of this tissue among specialists on flower structure. In many fruit types the pericarp can be separated into three regions: the outer exocarp, the central mesocarp, and the inner endocarp (see Table 5.1). These have diverse characteristics that contribute to the nature of the fruit. For example, in a drupe (e.g., the peach), the surface skin is the exocarp, the fleshy edible portion is the mesocarp, and the bony covering of the seed (stone) is the endocarp. In the coconut the outer skin of the fruit is the exocarp, the thick fibrous layer inside it is the mesocarp, and the bony inner shell is the endocarp (the edible part of the coconut is in fact the solid endosperm).

The nature of the fruit covering serves to protect the seed or seeds inside and may also contribute to seed dispersal. For example, some fleshy fruits are eaten by animals and the seeds are passed unharmed through the digestive system. In other cases single fruits may have wings for wind dispersal, or spines that attach to animal fur and are carried away, as in the beggar tick. Others may have buoyancy that allows transport by water (e.g., the coconut). A wide range of dispersal devices have been important in the evolutionary success of the angiosperms.

6

Shoot Morphology and Development

⁂ ───

THE SHOOTS OF vascular plants assume many different forms, but there is a pattern that is considered to be basic or fundamental. This is the upright or orthotropic leafy shoot, which consists of a vertical axis or stem bearing leaves at regular intervals along it. The stem is, in fact, a jointed structure in which nodes, the sites of leaf attachment, are separated by internodes, and this nodal/internodal organization is more than a superficial pattern as it is reflected in the internal structure of the stem (see Chapter 8). The shoot is viewed as consisting of repetitive units or modules, called phytomeres. Each phytomere consists of a leaf attached at a node, an internode, and an axillary bud. There is often a sharp zone of separation between leaf and stem in cases where leaves are shed at the end of their functional life, but during the active life of the leaf it is difficult to demarcate exactly where the leaf ends and the stem begins.

There is substantial variation in the extent to which the internodes are extended—that is, the degree of separation of successive leaves. Typically they are well spaced by elongated internodes in what is called a long shoot. Alternatively the internodes may be short so that the leaves are almost in contact, and the result is a short shoot. There are, of course, varying degrees of leaf crowding. Some plants are characterized by having exclusively long or short shoots. In others both occur, often with the short shoots borne as laterals on long shoots. In many instances a short shoot may change into a long shoot as it grows, or vice

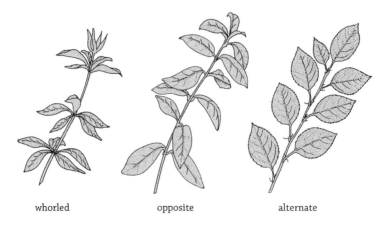

whorled　　　　　　　opposite　　　　　　　alternate

FIGURE 6.1 Different patterns of leaf arrangement (phyllotaxy) on a shoot: whorled, opposite, and alternate.

versa. The developmental basis of the long shoot/short shoot organization will be explained later.

Even a cursory examination of a leafy shoot reveals that the leaves are distributed in a regular pattern. The arrangement of leaves along the stem is called phyllotaxy, and there is a variety of phylotactic patterns. If two leaves are attached at a node and the leaves are placed 180° apart, the phyllotaxy is opposite (Fig. 6.1). If successive pairs of leaves occur at right angles to one another, the arrangement is further designated decussate. If three or more leaves are found at a node, the phyllotaxy is whorled (see Fig. 6.1), and the number in each whorl may be very large. When the leaves stand singly at the nodes and they are arranged in an ascending helix along the axis, the phyllotaxy is called helical or spiral. The term *alternate* is used only when a successive leaf occurs 180° around the stem from the preceding leaf (i.e., the leaves alternate on opposite sides of the stem; see Fig. 6.1). If one traces the successive leaves on the shoot axis the helical pathway is known as the generative spiral, and for any given species there is an equal probability that the direction will be clockwise or counterclockwise in any individual plant. The usual way to describe the phyllotaxy is in the form of a fraction in which the numerator is the number of times the stem is circled in proceeding from one leaf to the leaf vertically above it, and the denominator is the number of leaves passed through in this progression, not counting the beginning leaf. For example, if you begin to count at a particular leaf and pass five leaves in reaching one directly above, encircling the stem twice, the phyllotaxy is 2÷5, a very common phyllotactic pattern. In Figure 6.2, the phyllotaxy on left is 2÷3 and the one on right is 3÷8.

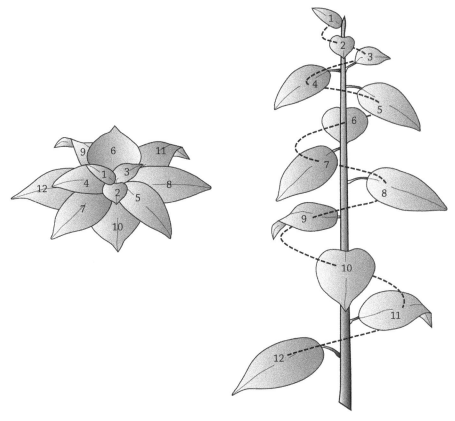

FIGURE 6.2 A diagrammatic presentation of 2+3 phyllotaxy (*left*) and a spiral arrangement of leaves on a branch with 3+8 phyllotaxy (*right*).

GROWTH OF THE SHOOT

As has been pointed out, it is a basic characteristic of shoots and roots of vascular plants that they remain meristematic or embryonic at their tips and are thus potentially capable of unlimited growth. As shown in the previous chapter, in both monocotyledons and dicotyledons the shoot and root apical meristems (SAM and RAM) are established during the development of the embryo and are present in a dormant condition until the seed germinates. In most species, during germination the root emerges first from the seed, followed by the shoot. The nutrient supply for the growing plantlet is derived from the endosperm or, as in the legumes and a number of other dicotyledon species, it may be absorbed from the cotyledons. The early seedling growth is varied among species; in some cases the cotyledons are carried above ground by the elongation of the hypocotyl, as in the bean plant (epigeal germination), or the hypocotyl may not elongate and the next internode with the shoot apex, the epicotyl, may elongate, leaving the cotyledons below ground (hypogeal germination), as in the pea. In the grasses, the single cotyledon, called the scutellum, remains adpressed to the endosperm and the coleoptile with the first leaf elongates, and the mesocotyl (the internode) may or may not elongate, depending upon the depth at which the grain is buried.

SHOOT APEX AND THE SHOOT APICAL MERISTEM

The meristematic region at the shoot apex is the SAM, and it gives rise to both the tissues of the stem and a succession of leaves in a repetitive fashion. The general organization of the shoot apex may be observed by carefully removing progressively younger leaves and leaf primordia in succession with a needle or fine forceps. When this operation is carried out under suitable magnification, several features of importance may be noted. The first is that the phyllotactic pattern (spiral or helical phyllotaxy) can be followed almost to the summit of the shoot (Fig. 6.3). The pattern is genetically controlled and is established at early stages, and leaf primordia are initiated as tiny outgrowths in precise locations at the margin of a central area that is free of outgrowths. This central region is the SAM, the generating center of the shoot, and it commonly has the form of a dome (see Fig. 3.9a in Chapter 3 and Fig. 6.3) or it may be relatively flat (Fig. 6.4) or an elongated cone. In the region around and just below the SAM the leaf primordia and young leaves are very close together (see Fig. 6.3)—that is, the internodes are very short.

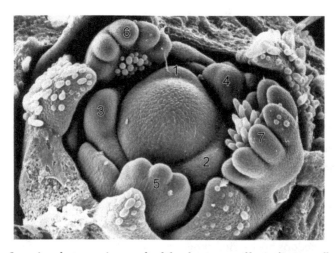

FIGURE 6.3 Scanning electron micrograph of the shoot apex of lupin (*Lupinus albus*) showing the apical dome and the spiral (helical) pattern of leaf primordial arrangement, numbered 1 to 7, around the dome.

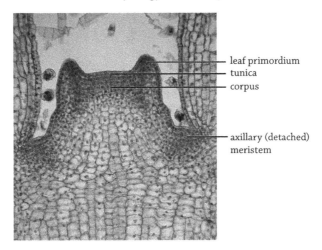

FIGURE 6.4 A median longitudinal section of the shoot apex of *Coleus*. Note the tunica–corpus organization of the SAM, the initiation of leaf primordia, and location of axillary (detached) meristem. (Courtesy of Dr. Edward Yeung)

The important role played by the SAM in giving rise to the shoot system necessitates that this region be examined in greater detail at the level of cellular structure. This is best done by studying sections, and those cut longitudinally are the most revealing (see Fig. 6.4). In the flowering plants it is possible to recognize one or more cell layers over the surface of the shoot apex in which cells divide almost exclusively in the anticlinal plane—in other words, new cell walls are formed at right angles to the surface. The opposite plane is the periclinal, in which new walls are formed parallel to the surface. This surface region, which is generally two or three layers but could be up to five layers, is called the tunica, and it covers a central cell mass, the corpus, in which divisions occur in all planes (see Fig. 6.4). The terms *tunica* and *corpus* are commonly used in descriptions of angiosperm SAM, and the number of tunica layers is usually specified. The origin of leaf primordia at the margin of the SAM is also seen in such sections, and it is clear that they are initiated by cell division activity in several layers of the meristem. In describing such a shoot apex the tunica layers refer only to the meristem above or within the youngest leaf primordia.

In addition to the tunica–corpus pattern of flowering plant shoot apices, there is another organization that may be recognized, more clearly in some cases than in others. This is a radial zonation in which a central zone may be distinguished from an encircling peripheral zone at the margin of which leaf primordia are formed (Fig. 6.5). Where clearly distinguished, the central zone consists of cells that are larger and less densely stained in histological preparations than those of the peripheral region. This pattern is not an alternative to the tunica–corpus organization but is superimposed upon it. Careful analyses have demonstrated

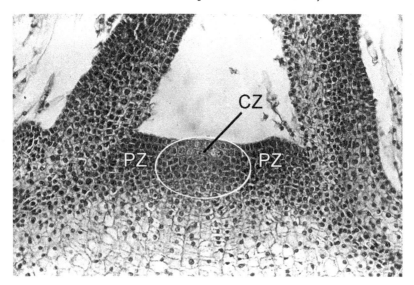

FIGURE 6.5 A median longitudinal section of shoot apex of sunflower (*Helianthus annuus*) with a central zone (CZ) of relatively large cells surrounded by the peripheral zones (PZ) of smaller cells with dense cytoplasm.

that the cells of the central zone divide less frequently than those of the surrounding region, in fact two to three times less frequently, and the cytological characteristics seem to reflect this reduced level of activity. However, the lower division frequency of the central zone is also found in cases in which the cytological features are not evident, and it seems to be a fundamental property of the SAM. This conclusion is strengthened by the fact that a comparable zonation pattern is found in gymnosperms, or nonflowering seed plants, where the tunica–corpus organization does not occur. In fact, it is the dominant characteristic of gymnosperm shoot apices. In many of the less advanced or non–seed-bearing vascular plants, such as horsetails and ferns, the meristem is organized around a distinctive apical cell in its center, and anticlinal divisions of this cell continually replenish the surface layer that gives rise to the shoot.

It is evident that the shoot apex must contain a permanent or self-perpetuating meristem, designated the promeristem; this is located at the extreme tip above and between the most recently formed leaf primordia (Fig. 6.6). It is equally clear that as growth occurs by cell division and cell enlargement, these cells must become specialized to function in the mature shoot. This occurs in the process of differentiation, which was introduced in the embryo in Chapter 5. Cellular specialization begins immediately behind the promeristem in the region of initial differentiation. As in the embryo, initial differentiation consists of the blocking out of the three tissue systems as protoderm, provascular tissue and the

Shoot Morphology and Development

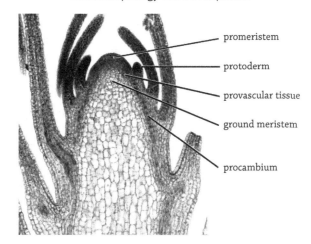

FIGURE 6.6 A median longitudinal section of flax (*Linum usitatissimum*) shoot tip showing the promeristem and the differentiation of protoderm, provascular tissue, procambium, and ground meristem.

procambium, and ground meristem (see Fig. 6.6). Subsequent differentiation that converts these into the mature tissues of the stem will be considered in Chapter 8. At the same time the leaf primordia initiated at the margin of the promeristem undergo growth and differentiation.

Because differentiation is a gradual process, there is no sharp boundary between the promeristem and the region of initial differentiation, but it is important to understand the difference between them. The fundamental property of the promeristem is that its cells retain the capacity to divide indefinitely, thus providing for the potentially unlimited growth of the shoot. This does not mean that the promeristem itself produces all, or even most, of the cells that compose the shoot. Rather, derivatives of the promeristem also multiply in the region of initial differentiation and in the maturing regions of the stem as well as in the leaves. The difference is that although the cells have begun to differentiate and divide actively, they will ultimately mature and cease to divide, with the exception of those that give rise to secondary meristems (see Chapter 11). It is the perpetuation of cell division rather than its quantity that enables the promeristem to play its essential role. There is reason to believe that the low mitotic frequency in the central zone of the promeristem may be necessary for the continuation of this activity.

SHOOT EXPANSION

In the shoot apex the leaf primordia are initiated very close together vertically and are not separated by extended internodes. In a typical long shoot (i.e., if the

FIGURE 6.7 A terminal bud of Ash (*Fraxinus* sp.) with protective cataphylls (bud scales) and its growth in the spring with expanding leaves and internodes.

leaves are well spaced by elongated internodes), the expansion occurs behind or below the shoot apex. This process is strikingly illustrated by the opening of the winter buds of woody plants in the spring. Dissection of the dormant or resting bud reveals a succession of immature leaves and leaf primordia closely clustered around the terminal meristem. As growth resumes the preformed leaves expand and the subtending internodes elongate at the same time, forming the new season's growth increment (Fig. 6.7). Additional leaf primordia may or may not be initiated and expand during the current season, but in either event, later in the season, a new bud is formed for the next growth period.

Thus it is evident that, while the terminal SAM is the essential generating center, most of the growth of the shoot (i.e., the increase in size) occurs outside the actual meristem. If this internodal elongation is greatly restricted, the result is a short shoot, although the leaves expand and an increase in diameter occurs. In many woody plants the resting bud is protected at the outside by a series of modified leaves or cataphylls or bud scales (see Fig. 6.7). Commonly there is little or no elongation of the internodes that subtend these scales, and their scars form a girdle that marks the beginning of the annual increment. In most non-woody plants the expansion of the shoot is not restricted to a limited period but goes on continuously as the new phytomeres are formed by the meristem. Thus this phase of development is less obvious than in woody plants, but it is not different in nature.

As stated above, much of the mitotic activity that produces the cells of the shoot occurs outside the SAM, but it is cell enlargement together with cell division that is involved in shoot expansion and the growth of the internodes. Both cell division and cell enlargement always participate, but the relative proportions of the two differ in different species. Cell division alone will result in a large number of smaller cells, but there will be no net growth. In order for growth to occur, cell enlargement must take place before, during, or after cell division, and the extent of it may vary. It may also be observed that each internode functions as a unit in the growth process, the growth rate rising to a maximum and then declining to zero. In some cases most of the growth of the shoot may be concentrated in one internode at a time successively along the stem, but in others several internodes are elongating at the same time. Commonly growth ceases first at the base of each internode and the cessation progresses toward the apical end. In this progression cell division stops first and the final phase of growth is accomplished by cell enlargement. In some plants, in particular many monocotyledons such as grasses, the cessation of growth occurs first at the top of the internode and advances toward the base. Such plants often show a distinctly jointed stem structure, and this tendency is enhanced in cases in which a growth zone persists for some time at the base of the internode. Such a growth zone with mature tissues both above and below is called an intercalary meristem. Plants having this feature often display a remarkable ability of seemingly mature shoots to right themselves after being forced out of the normal vertical position by unequal growth in the intercalary meristem.

SHOOT BRANCHING

In some plants an original axis remains unbranched, for example some palms, but it is much more common for branches to be formed. Two fundamental types of branching are recognized, terminal and lateral. Terminal branching, which occurs in many less advanced vascular plants but only rarely in seed plants, results from the subdivision of the terminal SAM into two equal or unequal growth centers that give rise to two equal (e.g., in *Lycopodium*) or unequal (e.g., in *Selaginella*) branch shoots,. Much more commonly, and overwhelmingly in seed plants, branches arise as laterally placed buds usually in the axils of leaves—that is, in the angle between the leaf and the stem upon which it is situated. These buds generally have their origin from small pockets of meristem derived from the promeristem of the main axis and left behind in the axils of leaf primordia without undergoing differentiation. These meristematic pockets, called detached meristems or axillary meristems (see Fig. 6.4), subsequently enlarge, begin to initiate leaf primordia, and form lateral buds. Sometimes it is difficult to trace the origin of the detached meristems from the terminal meristem, and it is reported in such cases that cells that have partially differentiated return to the full meristematic state before giving rise to a bud.

An axillary bud, once initiated, usually remains in an arrested state until it is some distance removed from the terminal shoot apex before developing as a branch. Indeed, in some cases it may not grow at all unless the terminal apex is removed. This is a manifestation of the strong control that the terminal apex appears to exert over lateral branching. This phenomenon is called apical dominance and is, in part, regulated by plant hormones; auxins produced in the shoot tip inhibit and cytokinins in the root tip promote growth of buds. Often the growth of lateral buds is markedly less vigorous than that of the main axis, an extreme example being the development of lateral short shoots on a main axis that is a long shoot. Another form of control is seen in the development of horizontally flattened lateral branches, as in many conifers. The diverse patterns of lateral branch development contribute significantly to the overall form or the architecture of the shoot system. In Chapter 4 it was noted that shoot apices may arise on a variety of plant structures either naturally or under artificially imposed conditions, often through the reactivation of mature tissues. Buds or shoots that arise in this way are called adventitious, and they may occur on a shoot in other than axillary positions. Usually their occurrence is in response to some sort of wounding or injury, and they result in branching additional to that normally found.

SHOOT MODIFICATIONS

Up to this point various aspects of the morphology and development of typical leafy shoots have been considered. There are, however, many modifications of this basic pattern that relate to either natural or artificial vegetative reproduction, and some of these have been considered in Chapter 4. Others, however, are concerned with the growth habit or the environment of the plant. What follows is a brief account of some of the more commonly occurring shoot modifications.

In some plants the leaves are greatly reduced and the photosynthetic function is performed largely by the stem. Such stems may become broad and flattened and may actually assume a form resembling a foliage leaf. These leaf-like stems, called cladophylls, as in the cactus *Opuntia*, may be identified by their position in the axils of reduced scale leaves, which demonstrates that they are lateral shoots. In many plants adapted to arid conditions or to habitats of high salinity the leaves are similarly reduced and the photosynthetic stems become very fleshy

Shoot Morphology and Development   67

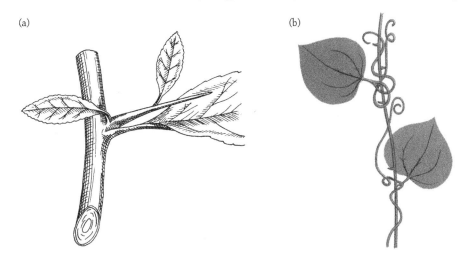

FIGURE 6.8 Modified shoot types: (*a*) thorn and (*b*) tendril.
a from *Botany: An Introduction to Plant Biology* by T. E. Weier, C. R. Stocking, and M. G. Barbour, 5th edition (1974), with permission of Dr. M. G. Barbour.

or succulent through the development of water storage tissue. Extreme expressions of this morphology are seen in the pad and barrel forms of cacti and other desert plants.

Two highly modified shoot types are the presumably protective thorns (Fig. 6.8a) and the tendrils (see Fig. 6.8b) that attach vines to a structure for support. In the development of a thorn, which may be an entire branch or just its tip, the shoot apex after producing a few reduced leaves becomes elongated and is converted into a hardened pointed tip as growth ceases, as in *Citrus*. Thorns may be unbranched or may bear laterals that become hard and pointed in the same way. Most tendrils are modified leaves or leaf parts, but some, as in the grape family (Vitaceae), are long slender stems that may produce branches in the axils of reduced scale leaves. They attach the parent shoot by curling around a support structure as a result of unequal growth stimulated by contact.

REPRODUCTIVE SHOOTS

One modification of the shoot that requires special attention because of its functional significance is the flower, which has been considered in detail in Chapter 3. A flower develops through the transformation of a vegetative shoot apex into a floral apex, often in response to a specific environmental stimulus such as altered day length. Typically the shoot apex enlarges, becoming either broad and flattened or elevated and dome-like. This change in shape is associated with accelerated mitotic activity throughout the SAM, including the previously less active central zone. Although the tunica and corpus continue to be identifiable, at least for a time, the meristematic activity tends to become more superficial in a mantle extending over a core of enlarging and less meristematic cells. This is associated with a heightened rate of floral organ primordia initiation, and the regular sequence of organ formation is usually altered. Often the sepals and petals are initiated sequentially in rapid succession but the stamens and carpels tend to be formed in simultaneous whorls. The regular acropetal (toward the apex) sequence of organ formation may be interrupted by the appearance of whorls basipetally to those already present. Ultimately the meristem is consumed in organ formation as the gynoecium is initiated and apical growth ceases (i.e., the shoot tip becomes determinate). The loss of indeterminate growth is associated with a uniform and high mitotic frequency throughout the meristem in contrast to the zonal organization of the vegetative SAM.

Flowers are often produced in inflorescences or flowering shoots, which have a variety of forms. Some are indeterminate and bear a succession of lateral flowers in axillary positions; others are determinate and may be highly modified, as in the head of a sunflower, which superficially resembles a flower (see Chapter 3 for details). In many cases the development of an inflorescence begins much as does the formation of a flower with a transformation of a vegetative apex (see Fig. 3.9). This is especially true in the case of determinate inflorescences in which the loss of vegetative zonation is associated with the onset of the determinate growth pattern. When an inflorescence has undergone such a transformation the individual flowers are often initiated as floral meristems directly without a prior vegetative phase. This evidence from determinate reproductive shoots lends support to the concept that the low mitotic frequency in the central zone of a zoned SAM is a necessary condition for the continuation of meristematic activity.

7

Plant Cells and Tissues

THE VASCULAR PLANT body is composed of a number of different kinds of cells, and these are associated in aggregates called tissues that perform the functions necessary for the plant's growth and survival. The patterns in which these tissues are organized differ in stem, leaf, and root, and these characteristic patterns will be examined in the following three chapters. There are, however, some fundamental features that are applicable to all structures of the plant body, and prior consideration of these will facilitate the detailed study of the individual organs.

The internal organization of all parts of the plant body can best be understood in terms of three tissue systems, the dermal, vascular, and fundamental or ground systems. These have already been encountered in the early differentiation of the embryo (see Chapter 5) and in the region of initial differentiation in the shoot apex as protoderm, provascular tissue and procambium, and ground meristem (see Chapter 6). Their appearance in the initial stage of differentiation is an indication of their significance in the organization of the plant body, as is the fact that they occur in all organs and encompass both the primary and secondary bodies. In broad terms also they have functional significance. The dermal system, which consists of the epidermis in the primary body and the periderm in the secondary body, provides protection from the external environment, particularly against excessive water loss and invasion of pathogens. The vascular system, which is composed of the two tissues xylem and phloem, is the conducting apparatus for water and inorganic and organic solutes and also offers mechanical support. The fundamental or ground system comprises the remaining tissues and is concerned with metabolic activities as well as support in many cases.

The basic structure of the plant cell has been presented in Chapter 2. Before considering the tissue systems in greater detail it will be helpful to examine the diverse cell types that result from the process of differentiation. There are a number of cell types and there is substantial diversity within each type.

CELL TYPES AND TISSUES

Within the mature or differentiated plant body a diversity of expressions of the basic cell pattern is found. In fact, the number of cell types is relatively small, but there are intermediate forms that complicate the picture. Certain cell types may have a scattered or dispersed distribution, but more commonly they are aggregated to form tissues. If a tissue is simple (i.e., composed of one cell type) it is designated by the same name as that of the cell type. Thus parenchyma is both a cell type and a tissue. Other tissues, such as xylem and phloem, are complex and are composed of several cell types arranged in distinctive patterns. An all-inclusive classification of tissues is difficult to devise and is not particularly helpful in understanding plant structure. Rather, it is more instructive to use tissues as a convenience in describing structural patterns. On the other hand, the tissue systems as previously defined provide a sound basis for the interpretation of all parts of the plant.

Parenchyma

Parenchyma cells could be called the basic cell type—that is, the least modified in the processes of differentiation (Fig. 7.1). They are typically isodiametric polyhedrons, but there are many deviations from this shape. They usually are surrounded by a thin primary wall, but in some cases there may be a secondary wall with simple pits. Parenchyma cells perform the major metabolic functions of the plant and accordingly retain a living, usually highly vacuolated, protoplast that varies with the function. Thus photosynthetic parenchyma, also called chlorenchyma, contains abundant well-developed chloroplasts. Storage parenchyma is packed with such metabolic products as starch grains, aleurone grains, fat droplets, or tannins. Some parenchyma cells release their accumulated products such as mucilages, resins, or essential oils, either by actual secretion through the plasma membrane and cell wall or by disintegration of the cell, and

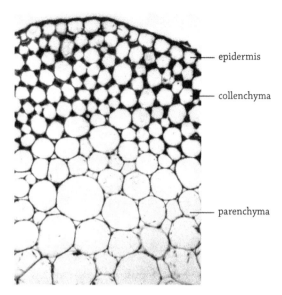

FIGURE 7.1 A cross-section of *Coleus* stem showing epidermis, collenchyma, and parenchyma.

are then designated secretory parenchyma. Parenchyma cells that are involved in the movement of solutes over short distances often develop extensive interior projections of the cell wall, thereby increasing the surface area of the plasma membrane and enhancing transport across the membrane. Such cells are called transfer cells.

Parenchyma cells may be aggregated in a tissue that is also known as parenchyma. This tissue is the basic component of the fundamental system, although other cell and tissue types are often included. On the other hand, parenchyma cells, either singly or in groups, may be distributed in other tissues such as xylem and phloem. One other type of parenchyma is recognized at the tissue level. When the intercellular spaces, which are characteristic of most parenchyma, are greatly enlarged, often as a consequence of particular cell shapes, the tissue is called aerenchyma and is believed to be important in the internal aeration of the plant and for the flotation of structures (e.g., the leaves of water lily; Fig. 7.2).

Collenchyma

In contrast to parenchyma, collenchyma cells are usually elongated in shape and they have primary cell walls, but the walls are thickened (see Fig. 7.1). The wall thickenings are uneven with different patterns, along the angles where the walls of adjacent cells meet (angular), on certain walls only (lamellar), or facing

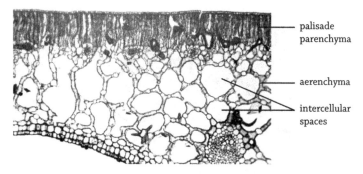

FIGURE 7.2 A cross-section of leaf of water lily (*Nymphaea*) with aerenchyma and large intercellular spaces.

intercellular spaces (lacunar). Because the wall thickenings are of primary wall, they are flexible and plastic so that the cells can grow and, while possessing considerable tensile strength, can bend without breaking. The high content of pectic substances gives the walls a glistening or "sticky" appearance in the living state and is responsible for the name (Greek *colla* = glue). Collenchyma cells are aggregated in strands or layers in the ground system, usually near the surface of a stem or leaf, in the petiole of a leaf (e.g., in celery), and rarely in roots. Their function is primarily mechanical in providing flexible support in growing regions and for herbaceous stems, where the ability to bend without breaking is important. Collenchyma cells often contain chloroplasts.

Sclerenchyma

Unlike collenchyma, sclerenchyma cells provide strong, rigid support both in the fundamental and vascular systems. Their strength is achieved through the deposition of a thick secondary wall with reduced inner cavity or lumen inside the primary and secondary walls (Fig. 7.3). The secondary wall is commonly impregnated with lignin, a complex polymer made up of aromatic alcohols. Most sclerenchyma cells are dead and devoid of a protoplast at maturity, but there are exceptions to this general feature. In aggregates they form a tissue also called sclerenchyma, but they may occur as isolated cells or small groups in other tissues.

Sclerenchyma cells are of two basic types, fibers and sclereids, the difference being largely one of shape. Fibers are elongated cells with pointed ends that overlap in aggregates. They are found in xylem and phloem but also form strands or layers in the fundamental system. Sclereids have a variety of shapes, ranging from isodiametric to columnar to bone-shaped with enlarged ends (see Figs. 7.3a and 7.3c), and are classified accordingly. They may occur singly, in clusters, or in

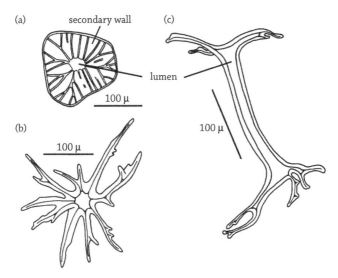

FIGURE 7.3 Various types of sclereids: (*a*) brachysclereid, (*b*) astrosclereid, (*c*) osteosclereid. From *Anatomy of Seed Plants* by K. Esau (1977), reprinted with permission of John Wiley & Sons Inc.

coherent layers as in seed coats. Particularly distinctive are the branched sclereids or astrosclereids (see Fig. 7.3b), which develop elongated arms or branches. If these branches are very long they may equal the length of fibers and the cell is called a trichosclereid. Sclereids of this type have a scattered distribution in other tissues, particularly the interior tissues of leaves. They are a typical example of idioblasts, individual cells that differ strikingly from those around them.

Tracheary Elements

The long-distance transport of water throughout the plant body occurs in two types of tracheary cells in the xylem, tracheids and vessel elements (Fig. 7.4). These cells resemble sclerenchyma in having a lignified secondary wall and are devoid of a protoplast when mature (i.e., are dead cells). However, the cell wall is not thickened to the same extent as sclerenchyma cells, leaving an adequate internal cavity through which the water moves from cell to cell. In effect they provide a channel of hollow tubes through which the water is pulled under tension in response to loss through transpiration or utilization. Since water also adheres to the cell walls, a major function of the secondary wall is the prevention of collapse of the cell in response to the internal tension. It is important to understand this function in examining patterns of secondary wall, particularly in the xylem of the primary body.

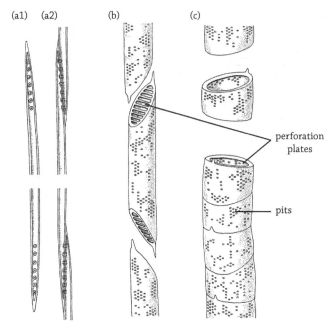

FIGURE 7.4 Three stages in the evolution of xylem: (*a1*) spindle-shaped tracheids of pine (*Pinus*) in which water is conducted through small-bordered pits on the lateral walls; (*a2*) tangential view of tracheids with overlapping pits, (*b*) vessel members in birch (*Betula*) in which water passes through partially dissolved end walls (perforation plates), (*c*) advanced vessel members in oak (*Quercus*) in which perforation plates are open and a series of vessel members are stacked one above the other to form a long tube, a vessel.

The secondary wall may consist of bands or strips in the form of separate rings (annular) or a helix (helical or spiral) (Fig. 7.5). There may be forks or anastomoses of the helix or more than one helix. A more extensive wall development leads to the scalariform (ladder-like) pattern, a network with slit-like openings, and if this is less regular the wall is reticulate. Alternatively the secondary wall is essentially complete, interrupted only by pits, and is called pitted, and these are of a characteristic type (see Fig. 7.5). Water passes easily from cell to cell through regions where there is only primary wall and the pit membrane has been modified but is impeded by lignified secondary wall. The bordered pits characteristic of some tracheary elements (Fig. 7.6a), generally in gymnosperms, provide a large exposure of primary wall necessary for water passage but reduce the weakening of the secondary wall. The pit membrane is overarched by a dome-like border of secondary wall with a relatively small opening (see Fig. 7.6b). This is particularly important where xylem contributes to the mechanical support of the plant, as in the secondary body.

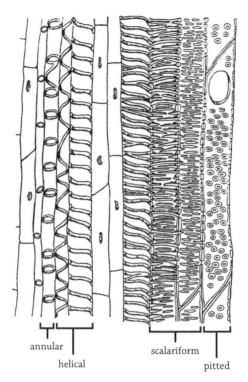

FIGURE 7.5 Tracheary elements in the stem of *Lobelia* with different patterns of secondary wall thickenings: annular, helical, scalariform, and pitted.

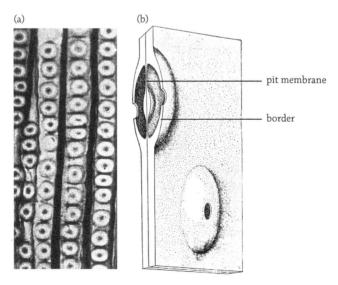

FIGURE 7.6 (*a*) Radial view of redwood (*Sequoia*) stem showing tracheids with circular bordered pits. (*b*) Diagram of a bordered pit structure showing pit membranes and an overarching border.

As noted above, tracheary cells are of two types, tracheids and vessel elements (see Fig. 7.4). Tracheids are elongated cells with overlapping pointed ends (see Figs. 7.4a1 and 7.4a2). The primary wall is intact and water passes through it from cell to cell where it is exposed in pits. Vessel elements are usually of larger diameter and are stacked end to end to form vessels (see Figs. 7.4b and 7.4c). In the cross-walls between adjacent vessel elements within a vessel, the primary wall is lacking, allowing water to pass without any impediment. These cross-walls are called perforation plates. Since vessels may be many feet in length, this modification is believed to increase the efficiency of water transport. Perforation plates may be perfectly transverse in orientation or they may be oblique to varying degrees. In some there is a complete opening, while in others they are crossed by transverse bands of secondary wall (see Figs. 7.4b and 7.4c). Tracheids are found in all groups of vascular plants and are considered to be the primitive (i.e., less advanced) or the original water-conducting elements.

Vessel elements have evolved from tracheids and are found only in angiosperms with just a few isolated occurrences in other groups. Tracheids and vessels occur in the complex tissue xylem, which also includes parenchyma cells and often fibers. Xylem in close association with phloem forms a continuous conducting system, the vascular system, throughout the plant body. The varied configurations of this system in stem, leaf, and root and in the secondary body will be an important consideration in the four following chapters.

Sieve Elements

The long-distance movement of dissolved organic substances is accomplished by specialized, elongated cells known as sieve elements or sieve tube elements (Fig. 7.7). In the angiosperms these are arranged in longitudinal files forming sieve tubes; this organization resembles that of vessel elements and vessels, but the sieve tube elements are totally different. They function as living cells rather than as empty conduits, but both the protoplast and the wall are very specialized. As a sieve tube element matures, its nucleus breaks down (i.e., the mature cell is lacking the nucleus) and the vacuolar membrane (tonoplast) disappears so that the distinction between cytoplasm and vacuole is lost. Mitochondria, plastids, and endoplasmic reticulum persist, but dictyosomes, ribosomes, and microtubules are absent at maturity. A particularly striking feature is the appearance of a unique filamentous protein known as p-protein. Sieve tube elements are extremely susceptible to disturbance and are distorted in most preparations.

The sieve tube elements have a primary cell wall but it may be somewhat thickened. These elements derive their name from the occurrence of sieve areas

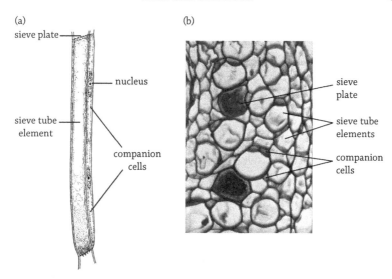

FIGURE 7.7 (a) Diagram of a sieve tube element with accompanying companion cells. (b) Cross-section of a stem of *Cucurbita* sp. through the phloem region with sieve tubes, sieve plates, and companion cells.

in the walls that are modified primary pit fields, and in which the plasmodesmata have enlarged to form intercellular protoplasmic connections. Sieve areas are particularly conspicuous on the end walls separating adjacent elements within a sieve tube. These end walls are called sieve plates (see Fig. 7.7), and the enlarged intercellular connections, through which the filaments of p-protein actually extend, are believed to enhance the longitudinal transport of materials along the sieve tube. Sieve plates may be inclined or horizontal and may contain several sieve areas or just one. The contrast between sieve plate, sieve areas, and lateral sieve areas on the side walls is usually striking.

The presence of a carbohydrate substance callose is a characteristic of sieve tube elements but is not restricted to them. Callose is composed of many β-1,3 glucose units, but it is not formed in microfibrils and it has no crystalline structure. Callose is associated with the sieve areas and plays a role in their formation by replacing the cell wall around the potential intercellular connections and then dissolving to permit their enlargement. In response to injury callose is synthesized rapidly on the sieve areas, effectively sealing them and preventing the escape of cell contents. A strong reaction of callose with the dye aniline blue is a useful indicator in the identification of sieve elements in histological preparations.

Sieve tubes are found only in the angiosperms; in the gymnosperms and seedless and less advanced vascular plants the transport of organic materials is carried out by a second type of sieve element, the sieve cell. Sieve cells like tracheids are long, tapered cells with overlapping ends. In most respects sieve cells resemble sieve tube elements, but they lack some of the specialized features such as p-protein. The most conspicuous difference from sieve tube elements is that there are no specialized sieve plates and sieve areas are distributed over the lateral walls.

Sieve elements are found only in the phloem, a complex tissue that, together with xylem, forms the continuous vascular system throughout the plant body. In addition to sieve elements phloem includes parenchyma cells and often fibers. In the phloem of angiosperms each sieve tube element is accompanied by one to several companion cells (see Fig. 7.7), small cells with a complete protoplast that arise as sister cells (i.e., from a common mother cell) of the sieve tube element. Companion cells are believed to function in the movement of materials into and from the sieve tube elements, which, lacking a nucleus, are relatively passive.

Epidermal Cells

Epidermal cells provide a protective covering for the entire plant body. Ordinary epidermal cells on the aerial parts of the plant are tabular in form with various shapes as observed in surface view. At maturity they retain a living protoplast but usually do not contain chloroplasts. Typically the outermost wall is thickened and the lateral and inner walls may be thickened to varying degrees. The outermost wall, and to some extent the other walls, are impregnated with cutin, a fatty substance, which impedes the passage of water. In addition to the cutinization of cell walls, a noncellular layer of cutin, the cuticle, is deposited on the outer surface and waxes are often deposited outside the cuticle. Thus the epidermis is well constructed to serve as a protection against desiccation and invasion of pathogens. In roots the cuticle is ordinarily very thin, especially in regions where absorption of water and minerals occurs. Epidermis usually consists of a single layer of cells tightly joined together without intercellular spaces. In some cases the epidermis is more than one cell layer in thickness, but only the outermost layer displays typical epidermal characteristics.

The structural features of epidermal cells that prevent water loss also interfere with the gas exchange necessary for photosynthesis. This problem is overcome by the presence of stomata (singular, stoma) in the epidermis of stem and leaves. A stoma is an opening in the epidermis bounded by a pair of guard cells (Fig. 7.8), which, by changes in shape in response to changes in their turgor, control the size of the opening and may close it completely. Unlike most epidermal cells, guard cells ordinarily contain chloroplasts. The guard cells are most commonly

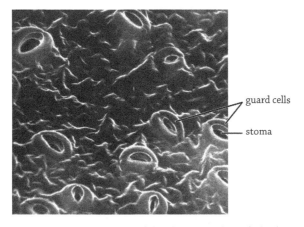

FIGURE 7.8 Scanning electron micrograph of the abaxial surface of a leaf showing epidermal cells with guard cells and stomata.

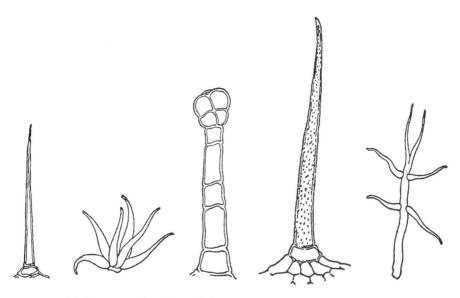

FIGURE 7.9 Various types of epidermal hairs.
From *Plant Anatomy*, 4th edition (1990), by A. Fahn, with permission of Elsevier Ltd.

crescent-shaped, although other forms occur, and differential thickening of the walls determines the shape changes. Thus gas exchange can occur under favorable conditions but water loss is controlled. Other specialized epidermal cells secrete a variety of substances, including mucilages, resins, essential oils, and nectar. Hairs or trichomes are a common feature of the epidermis of stems and leaves and are varied in structure. They may be unicellular or multicellular, often are branched in structure, and may also be secretory (Fig. 7.9). Finally, individual sclerenchyma cells may occur in the epidermis and, particularly in seed coats, the entire epidermis may be replaced by a layer of sclereids.

Secretory Structures

Secretion is a widespread phenomenon in plants, and there is a great diversity in the secretion products and the structures that produce and release them. Secretory parenchyma and secretory epidermis and epidermal hairs have already been mentioned. Secretory structures may be external, for example nectaries in flowers and the digestive glands of insectivorous plants, or internal, as in oil and resin ducts. Among the most distinctive of the internal secretory structures are the laticifers, unicellular or multicellular tubes that produce a complex material known as latex. Latex is often milky white in appearance but may also be clear or colored. In certain species it is the source of natural rubber. Multicellular laticifers are composed of superimposed cells in which the end walls break down, and they may form extensive systems in the plant body. Unicellular laticifers arise from individual cells that grow intrusively among the cells of other tissues, much like the hyphae of a fungus. They become multinucleate as they enlarge. In extreme cases laticifer primordia are differentiated only once in the embryo and their extension keeps pace with the plant as it grows.

Cork Cells

Cork cells are typically found in the periderm, the secondary dermal tissue, which replaces the epidermis as an outer protective layer as a stem or root enlarges. They are isodiametric or slightly elongated cells, often are somewhat compressed in the radial dimension, and are tightly packed without intercellular spaces (see Fig. 11.10 in Chapter 11). At maturity they are dead cells and their walls are impregnated with suberin, a fatty substance, which, like cutin, prevents the loss of water.

8

Tissues of the Stem

THE ORGANS OF the vegetative plant body, stem, leaf, and root are all constructed of the cell types and tissues described in the previous chapter, but each organ has its own particular pattern of organization. Moreover, there is considerable diversity in the expression of these patterns. This chapter and the two that follow will examine the structures of these organs and the developmental processes that give rise to them. The task of interpreting the diverse patterns and their origins is facilitated by the recognition that they all represent expressions of the three tissue systems, dermal, vascular, and fundamental or ground, which occur throughout the plant body and are the basis of its organization.

The stem is the axis of the shoot system and in a typical leafy shoot is a cylindrical, upright column bearing leaves at regular intervals and in a precise arrangement. As has been pointed out in Chapter 6, however, there are many specializations of the shoot system that depart from the typical form, and in such cases the stem is often highly modified. The structure of the stem reflects its two major functional roles, support of the shoot system and transport of water, minerals, and organic substances throughout the plant body. Other functions, such as photosynthesis or storage, are also performed and in modified shoots may be exaggerated at the expense of the more basic roles. The stem is a segmented structure in which nodes or zones of leaf attachment are separated by internodes, and the internal structures of nodes and internodes are distinctive.

In a long shoot the nodes are separated by extended internodes, whereas in short shoots they are crowded together and the internodes are of limited extent. The diameter of stems varies widely, from the slender axes of herbaceous annuals to the massive trunks of trees. This variation depends upon the extent of secondary growth, which ranges from none to many years of continued growth. Also, the shape of the stem varies; it is commonly round but can be somewhat flattened or angular. Although there are fundamental similarities between the tissues and their organization (anatomy) of dicotyledon and monocotyledon stems, there are also significant differences that require recognition.

THE DICOTYLEDONOUS STEM

The general arrangement of tissues may be observed in a transverse section of an internode of an herbaceous stem where the structures are largely or entirely derived from the shoot apical meristem (Fig. 8.1). At the periphery the external boundary is the single layer of epidermal cells that constitutes the dermal system. The vascular system has the form of a ring of bundles, each with an inner mass of xylem and an outer region of phloem; in some cases phloem may be found both inside and outside the xylem (described below). The bundles are separated by bands of interfascicular parenchyma. The cortex lies between the epidermis and the vascular tissue and a central core of pith is located inside the vascular tissue. The cortex and pith together with the interfascicular parenchyma make up the fundamental system or the ground tissue. The three tissue systems will now be examined in more detail.

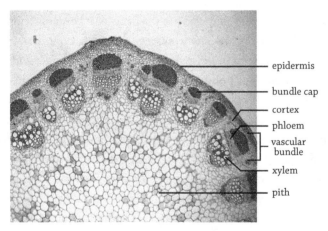

FIGURE 8.1 Transverse section of a portion of stem of sunflower (*Helianthus annuus*) to show the epidermis, cortex, vascular bundles (phloem and xylem), bundle cap, and pith.

The epidermis of the stem is composed of tightly joined epidermal cells, and stomata are usually present but often not very numerous. The entire surface is covered by cuticle, and in addition waxy deposits may occur on the cuticle. Hairs or trichomes may be present in the epidermis. The epidermis most commonly consists of a single cell layer, but there are instances of multiple epidermis in which the cells of the initiating layer divide periclinally to produce one or more inner layers. In such cases only the outermost layer is composed of typical epidermal cells. In cases where secondary growth occurs in the stem, the epidermis often keeps pace with stem enlargement by cell division and expansion.

The cortex or outer fundamental tissue is usually composed of chlorophyllous parenchyma with intercellular spaces. The stem cortex thus supplements the photosynthetic activity of the leaves and, where the latter are much reduced, may be the main photosynthetic tissue of the plant. Other cell types are also often found in various patterns. Collenchyma often is located just beneath the epidermis either as a continuous band of few to many layers or as strands. Fibers or sclereids may also occur as a subepidermal layer or may be distributed in other patterns in the cortex. A common pattern is the occurrence of clusters of fibers, called bundle caps (see Fig. 8.1), at the outer edge of the vascular bundles. Bundle caps, however, may be derived from the vascular tissue, and only a developmental study can determine the true origin. A distinctive layer of sclerenchyma or collenchyma just under the epidermis is designated the hypodermis. Another common feature of the cortex is the presence of secretory cells, either singly or in aggregates as secretory glands.

The pith is also composed predominately of parenchyma, but this is only rarely photosynthetic and has a mainly storage function. Prominent intercellular spaces are present, particularly in the central region, but the peripheral part may consist of more compactly arranged and smaller cells. Not infrequently the pith is torn out during the elongation of the internode, leaving a central cavity. This may be restricted to the internodes, leaving intact pith at the nodes. Conversely the pith may become sclerenchymatous, especially in the outer regions, through the lignification of the original parenchyma. True sclerenchyma in the form of sclereids may occur, and less frequently fibers are found adjacent to the vascular bundles. Secretory structures may also be present.

The vascular system has the form of a ring of vascular bundles or vascular strands of xylem and phloem located between the cortex and the pith. The xylem is adjacent to the pith and the phloem is peripheral to it next to the cortex (see Fig. 8.1). A vascular bundle with this arrangement is said to be collateral. However, it is not uncommon to find a second mass of phloem inside the xylem next to the pith, as in *Cucurbita* (Fig. 8.2). In such cases the bundles are designated

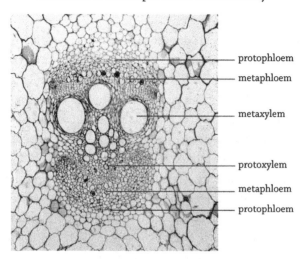

FIGURE 8.2 Bicollateral bundle in a stem of *Cucurbita* sp. with phloem both to the outside and inside of xylem.

bicollateral. The xylem contains tracheary elements, tracheids and vessels composed of vessel elements, and parenchyma. Fibers are sometimes present but are not common. The phloem consists of sieve tubes composed of sieve tube elements, companion cells, and parenchyma. Fibers are not usually present in active phloem but may develop later after the sieve tubes have ceased to function. These fibers may constitute the bundle caps, where they are not of cortical origin.

THREE-DIMENSIONAL ORGANIZATION OF THE VASCULAR SYSTEM

The major function of the vascular system is transport of materials to various parts of the plant body, and leaves are the main destination of water, obtained from roots, and are the major source of photosynthates transported through the stem. Thus a close association of the stem vascular system with the leaves is to be expected, and this is revealed if a section cut through the region of a node is examined. At each node one or more bundles or strands bend outward from the central ring and extend through the cortex toward the leaf as leaf traces (Fig. 8.3). Yet if the stem is examined at various levels along its length, the number of bundles in the ring does not change significantly. This indicates that bundles in the stem must branch—viewed in another way, as leaf traces enter the ring, they must at some lower level join with other bundles. These considerations lead to a three-dimensional picture of the vascular system, which is essential to an understanding of stem organization (see Fig. 8.3).

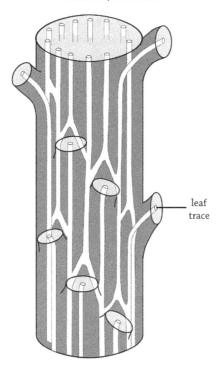

FIGURE 8.3 Three-dimensional diagram of the primary vascular system in a stem, with the pattern of leaf traces and their connections.

As each leaf trace enters the ring of vascular bundles in the stem it extends basipetally through several internodes before joining with another bundle adjacent to it. It may in fact branch and join more than one bundle. By convention a bundle from a leaf is considered to be a leaf trace in its basipetal extension as long as it remains distinct. Because, as has been pointed out in the discussion of phyllotaxy in Chapter 6, the leaves occur in vertical rows with precise intervals, the adjacent bundles should be those associated with leaves higher up on the stem in the same row. Thus the stem vascular system may be thought of as composed of leaf trace complexes equal in number to the number of vertical rows. The pattern of basipetal extension, that is the number of internodes and fusion with specific bundles, is extremely regular and is obviously closely related to the phyllotaxy. In some cases the leaf trace complexes remain entirely distinct along the length of the stem (an open system) or there may be links among the complexes so that the entire vascular system is interconnected (a closed system).

This description of the vascular pattern has assumed that each leaf is served by a single vascular strand, but often several traces are involved. These traces depart from the vascular ring at independent locations separated by continuing vascular bundles. There are always an odd number of traces (e.g., three or five), a median trace, and an equal number of lateral traces on each side. The lateral traces, like the median, have a regular pattern of basipetal extension and join in the stem bundle system. A further complication is that the median trace itself may consist of three or more bundles. Thus the stem vascular system may be quite complex, but the pattern is very regular.

The structure of nodes is described in terms of the pattern of leaf trace departures. When a leaf trace bends outward from the ring of bundles it leaves a temporary parenchyma-filled area confronting its point of departure, which is quickly obscured by the lateral displacement or branching of adjacent bundles above. Where there is a single trace consisting of one or several strands, the node is said to be unilacunar (the term *lacunar* refers to the absence of vascular tissue just above the trace departure). The very common three-trace pattern is designated trilacunar, and nodes with any larger number of departing traces are multilacunar. An important point to note is that this terminology refers to individual leaves; if there are two opposite leaves each with a single trace, the node is still considered to be unilacunar.

One additional factor contributes to the complexity of the stem vascular system. Commonly a leaf subtends an axillary branch whose vascular system must also be integrated into the system of the main axis. Typically in dicotyledons the branch vascular system enters the stem in the form of two traces, each consisting of one or more bundles, which are the traces of the first two leaves of the branch. These branch traces enter the stem bundle ring on either side of the leaf trace or the median trace if there are several, and extend basipetally in a definite pattern before joining adjacent bundles. Because their integration is associated with that of the leaf trace, they do not affect the nodal patterns described above.

DIFFERENTIATION OF TISSUES IN THE STEM

As discussed in Chapter 6, the tissues of the stem have their origin in the shoot apical meristem and begin to take form in the region of initial differentiation immediately subjacent to the self-perpetuating promeristem. In the region of initial differentiation small but significant differences in the frequency and orientation of cell divisions and in the pattern of cell enlargement begin to block out the three tissue systems (see Fig. 6.6 in Chapter 6 and Fig. 8.4). The single surface layer of protoderm marks the beginning of the dermal system. Internally a ring of provascular tissue (also referred as residual meristem by some authors) establishes the future vascular system (see Fig. 8.4). Ground meristem between

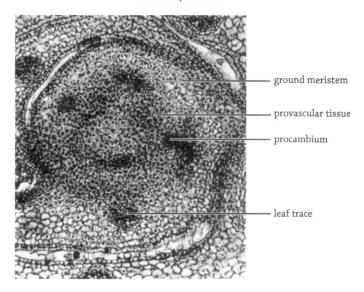

FIGURE 8.4 Transverse section of the stem of *Geum* (80 μM below the shoot apex) revealing a ring of provascular tissue in which procambial strands are differentiating in relation to leaf primordia.
From McArthur and Steeves (1972), *Botanical Gazette*, 133: 276–287, with permission of the University of Chicago Press.

the protoderm and the provascular tissue and in the center of the ring sets off the fundamental tissue system. Thus, the initial establishment of three submeristems, the protoderm, the ground meristem, and the provascular tissue, from which procambium will differentiate, sets at the very beginning the basic organization of tissues in the stem.

The subsequent differentiation of tissues may be traced by proceeding along the axis away from the shoot apex either in longitudinal sections (see Fig. 6.6 in Chapter 6) or in serial transverse sections. This is possible because the development of the shoot is indeterminate (i.e., it continues in a repetitive fashion) and all stages from initiation to final maturation are present in the stem at the same time. Differentiation in the root (see Chapter 10) may be studied in the same way, whereas in the leaf (see Chapter 9), which has a determinate developmental pattern, progressively older leaves must be examined to reveal stages of differentiation.

In the protoderm, cell divisions are restricted to the anticlinal plane so that it remains distinct from the internal tissues and ultimately differentiates into the epidermis. It is in fact a continuation of the outermost layer of the tunica in the promeristem. Proceeding away from the apex, the cells begin to vacuolate and enlarge while still continuing to divide as internodal elongation occurs (see Fig. 6.6 in Chapter 6). Growth of the epidermis continues as long as the stem elongates so that it remains an intact layer at maturity. Processes that convert protoderm cells into epidermal cells occur progressively, including thickening of cell walls, particularly the outermost wall, and secretion and deposition of cutin in the outer wall to form the cuticle and of waxes on the surface, where these occur. The development of specialized structures, including stomata and hairs, will be discussed later in relation to leaves where these processes have been more fully investigated.

The differentiation of the derivatives of ground meristem, both peripheral and central, is revealed in the early vacuolation and enlargement of the cells so that they appear less dense than the provascular tissue in stained sections (see Fig. 6.6 in Chapter 6). This differentiation is often more conspicuous in the potential pith region than in the presumptive cortex in the early stages. As internodal elongation occurs the cells divide predominately in the transverse plane, keeping pace with the growth of the stem and producing longitudinal rows of cells. There are also vertical divisions that increase the number of rows as the stem enlarges in diameter. As final maturation approaches, the cells enlarge and differentiate for the most part into parenchyma. Other cell types, collenchyma and sclerenchyma in the cortex and sclerenchyma in the pith, may also differentiate. In some cases, as noted earlier, final differentiation and the cessation of both cell division and enlargement occur in the pith before internodal elongation is complete and the tissue is torn out by continuing growth.

The early vacuolation of ground meristem leaves a ring of smaller and denser cells recognized as provascular tissue (see Fig. 8.4) or residual meristem (according to some authors). The two terms reflect a difference of interpretation of this tissue, whether it represents initial differentiation of the vascular system or merely meristematic tissue that has not begun to differentiate. In either case it is clear that this tissue blocks out the potential vascular region of the stem.

Within the provascular ring the differentiation of the tissue that will ultimately become xylem and phloem is recognized in strands of procambium and is associated with each leaf primordium as it emerges (see Fig. 8.4). Longitudinally oriented divisions produce cells that are longer than wide, although not necessarily longer than surrounding cells at the outset, and this characteristic shape is the identifying feature of procambium (see Fig. 6.6 in Chapter 6). The correlation of procambium differentiation with the leaf primordia is an early indication of the close association of mature vascular system with the leaves. The tissue of the ring that does not become procambium ultimately differentiates as interfascicular parenchyma.

Further differentiation of the vascular tissue may be observed by tracing the procambial strands basipetally (i.e., toward the base) along the axis, and there is complete continuity with mature vascular bundles. In other words, procambial differentiation is continuous and acropetal (i.e., toward the apex). An exception occurs in the branch traces of some axillary buds that are initiated after the ground meristem has undergone some differentiation. Then procambium may develop basipetally, establishing connections with the axial system. As differentiation proceeds the procambial cells undergo elongation meanwhile continuing to divide longitudinally so that the bundles increase in size. Thus the initial elongated shape is extended as the cells approach the condition of mature conducting elements. There is also an increase in the transverse dimension, particularly in those cells that will become xylem conducting elements.

The final differentiation of xylem and phloem involves a number of specific molecular and structural events that differ in the diverse cell types of these tissues described in Chapter 7. In the tracheary cells of the xylem, tracheids and vessel elements, lignified secondary wall is deposited in various patterns at an early stage and the perforation plates of the vessels are formed by dissolution of the primary walls and middle lamella. There is synthesis of a number of degrading enzymes in the cytoplasm, including DNAase, RNAase, and proteases, which are transported to the vacuole, leading to the collapse of the vacuole and membranous organelles (e.g., mitochondria, plastids, endoplasmic reticulum, and dictyosomes), and the complete loss of the protoplast. Thus, tracheary elements with thick secondary walls and empty spaces in the middle are formed, and are dead cells. Fiber differentiation is similar but with more extensive secondary wall deposition. Parenchyma is differentiated in strands derived by the transverse septation of procambial cells. In the phloem, sieve tube elements do not lose all their organelles. For example, mitochondria and plastids are retained, but the vacuole loses the tonoplast, the nucleus degenerates, and rough endoplasmic reticulum loses ribosomes and becomes smooth. The cell membrane is intact at maturity and the protoplast has special protein bodies called P-proteins. There is the development of sieve areas in the walls, most conspicuously the sieve plates in end walls. Companion cells with intact protoplasts are formed as sister cells of the sieve tube elements, and parenchyma and fibers differentiate as in the xylem.

An important aspect of these final differentiation processes is that they do not occur uniformly across each bundle. As a procambial bundle is followed basipetally in serial transverse sections, the first mature tracheary elements appear adjacent to the pith and the maturation appears to spread across the bundle toward the phloem region (see Figs. 8.1 and 8.2). The first mature xylem elements to appear are designated the protoxylem and those that differentiate subsequently as the metaxylem (see Fig. 8.2). This pattern of differentiation is centrifugal and is referred to as endarch because the first or oldest elements are located at the inside of the bundle. In the phloem region the first mature elements, the protophloem, occur adjacent to the cortex and metaphloem differentiation advances toward the xylem region (i.e., center of the stem or centripetally; see Fig. 8.2). There is no special term for this pattern in the phloem.

The first elements to differentiate are often conspicuously smaller than those that mature later. In the xylem there is also a sequence of secondary wall patterns in the successively differentiated tracheary elements beginning with annular or helical and progressing to more continuous scalariform or reticulate walls and ultimately complete walls interrupted only by pits (see Fig. 7.5 in Chapter 7). The significance of this sequence is evident when it is considered in relation to stem elongation. Annular and helical elements, although they cannot grow, can be stretched passively and thus serve as functional water-conducting elements in the elongating region of the stem. The stretching often results in the rupture of the early formed elements, sometimes leaving a recognizable cavity, but the sequential differentiation replaces them with new conducting cells so that the supply of water to the growing region is not interrupted. When elongation ceases, nonstretchable scalariform, reticulate, or pitted elements differentiate, completing the maturation of the xylem. A similar process occurs in the phloem region, although there are no correlated differences in wall patterns. Phloem matures in advance of the xylem in each bundle; indeed, protophloem elements are generally the first to mature in a bundle, and the differentiation of protoxylem generally does not begin until phloem maturation has reached that level.

The terms *protoxylem*, *metaxylem*, *protophloem*, and *metaphloem* are useful in identifying the location of the first maturation of conducting elements and describing the sequence of differentiation, but there is no generally accepted rule as to how much of the tissue should be allotted to each category. Some authorities, however, have adopted the convention of delimiting protoxylem and protophloem as the tissues that mature during elongation and metaxylem and metaphloem as those that mature after elongation ceases.

A further complication arises when the longitudinal course of xylem and phloem maturation is considered. Phloem, like the procambium, differentiates in a continuous pattern in an acropetal direction. Differentiation of the first xylem, the protoxylem, begins discontinuously near the base of each leaf primordium and advances in two directions, acropetally into the leaf trace and basipetally in the stem. The basipetally differentiating protoxylem ultimately joins mature xylem differentiating acropetally in the bundle.

Proceeding basipetally along a bundle, the maturing xylem and phloem advance toward each other from the inside and outside respectively. Meanwhile, longitudinal divisions continue in the procambium that separates them. As differentiation proceeds the dividing procambium is progressively restricted and in extreme herbaceous plants differentiates completely as mature xylem and phloem meet. In such cases there is no secondary growth and the stem vascular system is entirely primary. However, in many herbaceous plants and all woody species, the dividing zone persists and initiates the vascular cambium and secondary growth in the stem (for details see Chapter 11). Thus secondary xylem and phloem are added to the primary tissues of the bundle. If secondary growth is limited, it may be restricted to the bundles, but commonly the parenchyma of the interfascicular regions begins to divide and the vascular cambium becomes a continuous ring. The secondary tissue in the interfascicular regions may be similar to that found in the bundles, but often only parenchyma or fibers are formed rather than typical xylem and phloem (detailed in Chapter 11).

THE MONOCOTYLEDONOUS STEM

The three-tissue-system pattern is also found in the monocotyledonous stem, but the organization of the vascular system appears to be very different from that of the dicotyledons. The most striking difference is that the vascular bundles are not confined to a ring around the pith but rather are distributed through the ground or fundamental tissue (Fig. 8.5) or sometimes occur in two circles around the pith. The leaves of monocotyledons typically have a large number of leaf traces, which enter the stem by way of a sheathing basal region. Careful analysis has shown that the bundles in the stem are all related to leaf traces and follow a definite course along the axis. Each bundle periodically bends abruptly outward toward the periphery, branches to form a leaf trace, and then follows a course back toward the center. Branches connecting to other bundles also occur at this point. Vascular bundles have a range of sizes, the larger near the center and the smaller toward the periphery (see Fig. 8.5). The larger central bundles give rise to leaf traces less frequently than do the smaller peripheral ones. Thus the complex vascular system of the monocotyledons is not fundamentally different from that of the dicotyledons in spite of appearances. Because the vascular bundles are scattered in the stem, there is no clear distinction of the cortex from the pith of the ground tissue.

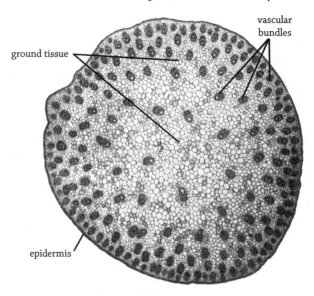

FIGURE 8.5 Transverse section of *Zea mays* stem (a monocot) showing the scattered distribution of vascular bundles and the various sizes (smaller to the outside and larger toward the center).

The basic pattern of tissue differentiation in a stem (dicot or monocot) as well as in a root is summarized in Figure 8.6, with the exception that in a monocot stem there is no differentiation of vascular cambium.

THE STELE

The organization of the vascular tissue of the stem as a system related to the leaves is found throughout the seed plants generally, but only in the stem. If one examines the roots (see Chapter 10) a very different pattern is found. In roots there is a central core of xylem, usually star-shaped in cross-section, with the phloem peripheral to it and located in the bays between the xylem arms. A similar vascular pattern is found in the stems of some lower vascular plants, including the most primitive found in the fossil record. Consideration of these facts has led to the concept that the primary vascular system is best regarded as a unit, however much it may be dissected in some cases. This unitary system, the central column of vascular tissue and associated fundamental tissue, is called the stele (column). The terminology widely used in describing vascular systems is based on the stelar concept (Fig. 8.7). A system with a central core of xylem surrounded by phloem observed in the stem of early vascular plants (e.g., *Psilotum*) is called

Tissues of the Stem

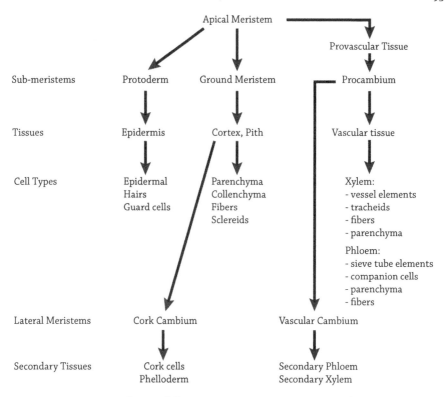

FIGURE 8.6 The pattern of tissue differentiation in a stem or root. Note there is no differentiation of vascular cambium and cork cambium, and the respective secondary tissues, in a monocot stem.

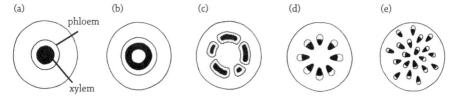

FIGURE 8.7 Different types of steles: (*a*) protostele, (*b*) siphonostele, (*c*) dictyostele, (*d*) eustele, (*e*) atactostele.

a protostele ("first stele," see Fig. 8.7a) and is regarded as the original type from which others have evolved, in stems but not in roots.

Some ferns also have a protostele, but most have central pith surrounded by a cylinder of vascular tissue. This is called a siphonostele (tubular stele, see Fig. 8.7b). The separation of each leaf trace is confronted by an area of fundamental tissue that interrupts the vascular cylinder, a leaf gap. In many ferns the leaf gaps are very large and overlapping so that the vascular tissue forms a network.

A stele of this type is called a dictyostele (net stele, see Fig. 8.7c) and may superficially resemble that of a seed plant but is fundamentally different in that the bundles of vascular tissue are not leaf traces or extensions of them. The leaf-related bundle system of seed plant stems, as in dicotyledons, has been named a eustele (true stele, see Fig. 8.7d); this is something of a misnomer since it shows little resemblance to a central column or stele. The system of the monocotyledons with scattered bundles is often designated an atactostele (stele without arrangement, see Fig. 8.7e) based on the misconception that there is no orderly arrangement in the bundle pattern.

9

The Leaf

―――――

LEAVES ARE LATERAL appendages borne on a stem, and they along with the stem make up the shoot. The shoot apical meristem (SAM), which initiates the tissues of the stem, also gives rise to a continuous sequence of leaf primordia in a regular pattern recognized in the phyllotaxy of the shoot (see Chapter 6), and the coordinated development of stem and leaf is reflected in the tissue differentiation in the stem (see Chapter 8). Unlike the shoot as a whole, which is indeterminate or potentially unlimited in growth (see Chapter 1), the leaf is determinate—that is, after a defined period of growth it matures completely. The extent of development, however, varies greatly, producing organs that range in size from tiny scales to the fronds of some palms, which are many meters in length. Leaves are also temporary organs that in long-lived plants are discarded after a time, a phenomenon called leaf abscission, which is readily observed in deciduous trees in regions with alternating seasons.

The leaves are the most important photosynthetic organs of the plant body, and the structure, both external and internal, of a typical foliage leaf reflects that function. A leaf usually has a broad, flattened structure—the leaf blade or the lamina, which exposes the maximum surface area to incident light. It is also dorsiventral—that is, it has distinct upper (adaxial) and lower (abaxial) surfaces and its symmetry is bilateral. In all of these features it contrasts markedly with the typical stem. Internally the tissue organization is related to both the light absorption and the gas-exchange functions required for photosynthesis.

There are many variations in the structure of foliage leaves, but these represent only a small fraction of the diversity of forms that leaves assume in relation to particular functions. Bud scales, spines, tendrils, storage organs, absorbing structures, and insect traps are only a few of the specialized foliar organs. In addition, the protective, attractive, and reproductive organs of the flower are also considered to be foliar in nature (see Chapter 3). Indeed, no other plant organ has such a wide range of forms. Here we shall deal mainly with foliage leaves, but this limitation allows ample scope for diversity, particularly as it is related to environmental factors.

LEAF FORM

The diversity of leaf form is enormous, so it seems wise to begin with a basic type represented by the foliage leaf of the dicotyledons. The typical leaf consists of an expanded lamina, a stalk or petiole (Fig. 9.1a), and a more or less distinctive leaf base where the petiole is attached to the stem at a node. However, in many instances the petiole is absent and the leaf is said to be sessile. The leaf base may be sheathing—that is, partly or completely encircling the nodal region of the stem (Fig. 9.2). There may also be a pair of stipules at the base of the petiole (see Fig. 9.1a) or on the stem at the junction, which are considered to be part of the leaf because their vascular supply connects to that of the leaf. Leaves are designated as simple when there is a single lamina and compound when more than one unit (leaflets) contributes to the laminar tissue of the leaf (see Figs. 9.1b and 9.1c).

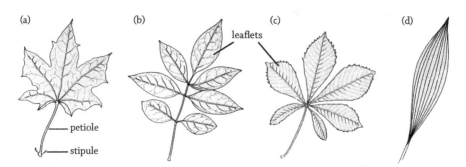

FIGURE 9.1 Different leaf types and venation patterns: (a) simple leaf with palmate venation, (b) pinnately compound leaf with pinnate venation, (c) palmately compound leaf with pinnate venation, (d) simple leaf with striate (parallel) venation.

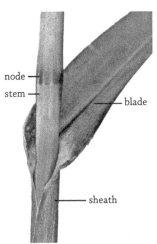

FIGURE 9.2 A grass leaf with a blade and a sheath encircling the stem below.
From *Botany: An Introduction to Plant Biology* by T. E. Weier, C. R. Stocking, and M. G. Barbour, 5th edition (1974), with permission of Dr. M. G. Barbour.

The lamina has a wide range of features based on overall shape, as well as the nature of the tip, margins, and base. These characteristics, which are important in plant classification and identification, may be relatively constant for a species or may vary to some extent with the stage of development or environmental conditions. The blade of a simple leaf may be lobed, in some cases very deeply, but in cases of compound leaves the lamina is clearly divided into separate leaflets (Figs. 9.1b and 9.1c) which may be associated with stipules. The leaflets may be arranged in two rows in a pinnate (feather-like) fashion along a rachis that is the extension of the petiole (see Fig. 9.1b) or may occur in a palmate (palm-like) pattern (see Fig. 9.1c) at the tip of a petiole. The morphology of the leaflets is similar to that of simple leaves. The leaflets of a compound leaf may be further compounded, and a threefold compounding is not unknown. In rare cases the two patterns of compounding may be combined, as in *Mimosa* (sensitive plant), which has palmately compound leaves with pinnately compound leaflets. There are often different leaf forms on the same plant (e.g., bud scales or tendrils as well as typical foliage leaves), a condition known as heterophylly. There may also be differences between juvenile and adult leaves, with different morphologies, and transitional stages between the two, a phenomenon referred to as heteroblasty.

In the monocotyledons the leaves most often have an encircling leaf base forming a distinct sheath (see Fig. 9.2), but this is not always the case. The grass leaf exhibits this condition in well-developed form where the blade or limb is attached directly to an extensive sheath that extends a considerable extent above the node at which the leaf is actually attached. In some monocots, for example the banana, there is a petiole between the blade and the sheath and the extensive leaf sheaths that encircle one another extend well above the shoot apex, forming a pseudostem. Among monocotyledons simple leaves are most common but compound leaves do occur. The complex leaves of many palms develop in a distinctive manner and will be discussed below.

Venation Patterns

One of the most distinctive features of leaf blades is the venation pattern (pattern of veins), which, although it refers to the vascular system, includes more than vascular tissue. In plants other than angiosperms the venation of the leaf, if it is more complex than an unbranched strand, has an open dichotomous pattern. This pattern occurs with extreme rarity in the flowering plants, where the venation ordinarily consists of an interconnected system that falls into one of two basic types, reticulate or net-like (Fig. 9.3a) and striate (see Figs. 9.1d and 9.3b), traditionally referred to as parallel. Studies carried out in recent years, however, have revealed much greater complexity and diversity in venation patterns and have led to a more detailed classification that is beyond the scope of this book.

Reticulate or net venation is characteristic of most dicotyledons and a few monocotyledons, for example the family Araceae. In this pattern a series of vein orders, up to five in number, in order of decreasing size are recognized, and the fine ultimate veinlets anastomose, forming areoles of nonvascular tissue within which there are often free vein endings (see Fig. 9.3a). In the pattern traditionally designated pinnate, a main vein or midrib extends from the base of the lamina to its tip, and from it smaller veins arise in a feather-like pattern (see Fig. 9.1b). Variations on this pattern depend upon the course of the smaller veins. Alternatively, in the pattern long known as palmate (see Fig. 9.1a), three or more main veins diverge from the base of the lamina toward the margin, giving rise to higher vein orders as they extend. Variations on this pattern in which two or more of the main veins or their major derivatives converge toward the leaf tip in particular ways have given rise to further categories. The elaboration of the classification of venation patterns has been of particular significance to taxonomists and to paleobotanists who often deal with isolated leaves.

Striate or parallel venation is characteristic of most monocotyledons but also of a few dicotyledons, for example *Plantago* (plantain). The superficial impression of striate venation is that a number of more or less equally spaced veins extend the length of the leaf in a parallel manner (see Fig. 9.1d). This character is associated with the typical sheathing leaf base of monocotyledons through which a large number of traces enter the nodal region independently. Detailed examination, however,

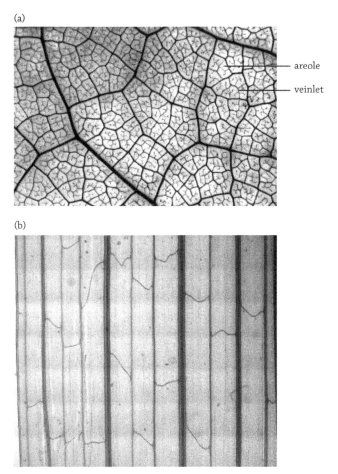

FIGURE 9.3 (*a*) A cleared leaf of *Liriodendron tulipifera* with reticulate venation and showing veinlets, vein endings, and areoles. (*b*) A portion of cleared leaf of barley (*Hordeum vulgare*) showing parallel venation with both major and minor veins and small veinlets interconnecting them.

has shown that veins do not necessarily extend the length of the leaf blade but that there often is convergence and anastomosing toward the tip. In other cases veins diverge from the median portion of the blade toward the margins and may then converge again. Furthermore, there are usually minor vein connections between the main longitudinal veins, a condition well illustrated in grass leaves, where smaller veinlets interconnect the more conspicuous longitudinal veins (see Fig. 9.3b).

Leaf Types

A detailed consideration of the many diverse leaf types is beyond the scope of this book, but several of the more common forms may be mentioned. The reader is referred to the illustrated volume by Bell (1993) for an account of this fascinating topic. Perennial plants that have seasonal growth often develop bud scales or cataphylls (see Fig. 6.7 in Chapter 6) around a dormant bud that protect the delicate foliage or floral primordia within from desiccation. These scales develop from primordia that, although initially equivalent to foliage primordia, follow a divergent developmental path from an early stage. Spines also develop from equivalent primordia but form slender, conical, and hardened projections lacking any lamina. Sensitive, twining tendrils that serve to attach a shoot to a support are another divergent leaf, and they may represent an entire leaf or only certain leaflets of a compound leaf (e.g., in *Pisum* or pea; Fig. 9.4). Leaves may also serve as storage organs, particularly as they are found in bulbs (e.g., in onion). Some of the most extreme leaf modifications are those that serve as traps for insects and other small animals, providing supplemental nutrients to the plant (i.e., plants eating animals). The trapping devices of these carnivorous plants are of several types, including little more than a sticky leaf surface, tentacles that respond to stimulation by enfolding the prey (e.g., in sundew [*Drosera* sp.]), a leaf blade with tentacles that folds hinge-like on the prey when stimulated (e.g., the Venus flytrap [*Dionaea* sp.]), and diverse types of vessels or pitchers into which the creature enters and is prevented from exiting (e.g., the pitcher plant [*Nepenthes* sp.]). In all cases digestive enzymes are secreted by the leaf blade that break down the animal, and the products of digestion are absorbed by the leaf.

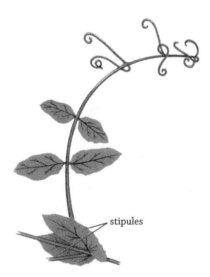

FIGURE 9.4 Diagram of a compound leaf of garden pea (*Pisum sativum*) with tendrils.

TISSUES OF THE LEAF

Petiole

The petiole of a simple or compound leaf and the rachis of a compound leaf superficially resemble a stem, but their tissue organization, while suggestive of that of a stem, is distinctive. The outermost layer of tissue is the epidermis, which resembles that of the stem, and the fundamental or ground tissue consists of chlorophyllous parenchyma, which is not separable into cortex and pith. It is very often supplemented by the presence of collenchyma, as in celery (*Apium* sp.), which provides flexible support, and less commonly by sclerenchyma. The vascular tissue consists of one to many strands or traces in continuity with the stem vascular system, which may branch or fuse along the extent of the petiole. The vascular strands may form an arc or a ring, may be irregularly disposed, or may be in the form of a ring with a bundle in the center. The bundles may be collateral with the xylem in an adaxial position if the stem bundles are of that type, but they may also be bicollateral or even concentric. At the base of the petiole in some species, a pad-like swelling called a pulvinus is present. In this structure the vascular tissue is grouped in the center with a large volume of parenchyma around it. The parenchyma cells expand or contract as a result of changes in turgor, and this provides a mechanism that can cause leaf movements. A similar structure may occur at the base of the leaflets of a compound leaf.

Leaf Blade (Lamina)

Although the lamina of a simple leaf or of a leaflet of a compound leaf has a very distinctive internal structure related to its role as the major photosynthetic organ of the plant, it may still be interpreted in terms of the three tissue systems (Fig. 9.5). The dermal system consists of an upper (adaxial) and lower (abaxial) epidermis, and epidermal cells are tightly joined together with the waterproofing substance cutin deposited on their outer walls (see Chapter 7). The noncellular cutin layer, the cuticle, is usually very well developed in leaves, and often waxes are deposited on top of the cuticle. The leaf is, therefore, well protected from excessive water loss except for the presence of stomata, which are essential for the gas exchange involved in photosynthesis. Stomata tend to be concentrated on the lower surface of the leaf blade but are not necessarily absent from the upper surface. A stoma consists of two guard cells and the space or pore between them (see Fig. 7.8 in Chapter 7) and whereas the epidermal cells generally lack chloroplasts, they are present in the guard cells. The pattern of wall thickening in the guard cells results

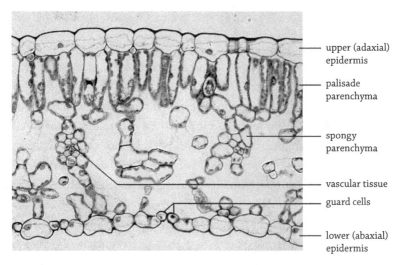

FIGURE 9.5 Transverse section of *Euphorbia lathyrus* leaf blade showing upper and lower epidermis, palisade and spongy mesophyll, and vascular tissue. (Courtesy of Dr. Riyadh Muhaidat)

in shape changes when turgor is increased or decreased (see Chapter 7) so that the stomata regulate water loss while allowing the entrance of carbon dioxide necessary for photosynthesis, and the exit of oxygen. The guard cells are often associated with subsidiary cells, as in corn leaf, which may or may not have a developmental relationship with the guard cells. The subsidiary cells are believed to participate in the opening and closing mechanism and together with the guard cells constitute the stomatal complex. The stomata may be at level with the other epidermal cells but often are sunken below that level, a condition believed to diminish water loss from the leaf, and there is often a sub-stomatal chamber below the guard cells. Hairs or trichomes are often present on the epidermis, and these may be unicellular or multicellular, branched or expanded into scales, and may be secretory.

The fundamental or the ground tissue of the leaf is known as the mesophyll and is the location of most or all of the photosynthesis carried on in the leaf. In most monocotyledons and some dicotyledons, especially succulents, the mesophyll consists of uniform parenchyma with intercellular spaces. However, typically in dicotyledons, but also in some monocotyledons such as *Lilium*, it is composed of two distinctive layers, the palisade and the spongy mesophyll or parenchyma (see Fig. 9.5). The palisade consists of cells elongated perpendicular to the leaf surface in one or more layers most commonly beneath the adaxial epidermis, but in some cases under the abaxial epidermis as well. In some cylindrical leaves it may occur all around the leaf. Most of the photosynthesis occurs in the palisade layer, which commonly contains 70% to 85% of the chloroplasts. The name *palisade* is somewhat misleading because it is not a compact layer but contains a great deal of intercellular space, as can be observed in sections parallel to the leaf surface. In the spongy mesophyll the cells are more or less isodiametric but tend to be somewhat irregular in shape and are interspersed by a large volume of intercellular space. This together with the spaces in the palisade layer results in an extensive internal atmosphere in the leaf, and the internal surface is many times that of the external surface. The significance of this organization is that it provides extensive moist surface for gas exchange without exposure to the external atmosphere and the resulting water loss.

The vascular system of the lamina has already been noted in the highly branched venation pattern (see Fig. 9.3). The veins contain more than just vascular tissue, but the total structure should be considered together. The major veins of a leaf commonly include one vascular strand, but there may be several. In these veins the actual vascular tissue is surrounded by parenchyma, often collenchyma and sometimes sclerenchyma in the form of fibers, and this interrupts the general pattern of palisade and spongy mesophyll. This aggregation of tissue frequently results in pronounced ribs that protrude on the abaxial face of the lamina, and this may apply to the larger minor veins in addition to the main veins. The relative positions of xylem and phloem found in the stem bundles, collateral or bicollateral, extend into the lamina veins so that if the pattern is collateral the xylem is above the phloem (i.e., adaxial to it). In some cases if there is a complex pattern in the petiole it may be extended into a major vein, particularly the midrib of a pinnately veined leaf.

Smaller veins contain less xylem and phloem and extend mainly through the spongy mesophyll. They are surrounded, however, by bundle sheath parenchyma (Fig. 9.6) so that they are not exposed to intercellular space. The bundle sheath cells are elongated and, in some cases where they are adjacent to the phloem they are of the transfer cell type, presumably related to the function of loading photosynthates into the conducting tissues. In other cases the sheath cells may have a Casparian strip similar to that of the endodermis in the root (see Chapter 10). At the termination of the vein branching system the ultimate vein endings may consist of xylem tracheids only. Commonly panels of cells of the sheath type extend from the bundle to the upper and lower epidermis layers in what are referred to as bundle sheath extensions (see Fig. 9.6) and are thought to transmit water from the bundle to surrounding cells. There may also be significant development of sclerenchyma associated with the vascular tissue and the bundle sheath extensions, especially in monocotyledons.

There is a particular type of bundle sheath found in some 19 families, including both dicotyledons and monocotyledons, and particularly well developed in some

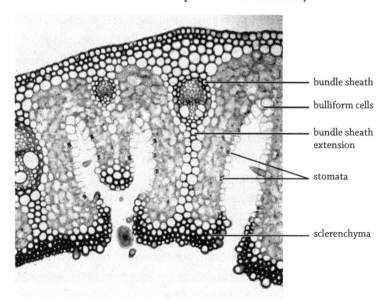

FIGURE 9.6 Transverse section of the leaf of *Ammophila*, a xerophyte, with bundle sheath, bundle sheath extension, and sunken stomata.

grasses. This bundle sheath specialization is significant because it increases the efficiency of photosynthesis at high temperatures and light intensities through a process known as C_4 photosynthesis. The cells of the bundle sheath are tightly held together and contain large chloroplasts without grana, whereas the ordinary mesophyll cells have small chloroplasts with grana. The cells of the bundle sheath form a radiating pattern that, along with a distinct chloroplast form, has been referred to as Kranz (wreath) anatomy. Oxygen inhibits photosynthesis and reduces its efficiency, particularly at high temperatures and light intensities through a process known as photorespiration. In C_4 plants, the mesophyll cells fix carbon dioxide into C_4 acids and pass these along to bundle sheath cells from which oxygen is largely excluded, carbon dioxide is released, and the C_3 photosynthetic reduction reactions occur.

VARIATIONS RELATED TO THE ENVIRONMENT

Many species show structural modifications of their leaves in response to particular environmental conditions such as drought or excess water, light intensity, and mineral deficiency. Leaves in bright sun (sun leaves) tend to be smaller and thicker compared to shade leaves and contain an increased number of palisade layers, a thick cuticle, and increased covering of hairs. In some sun leaves there

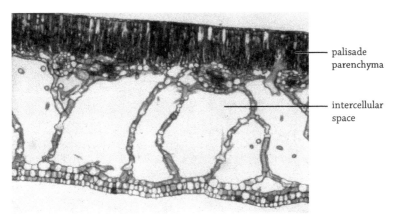

FIGURE 9.7 Transverse section of the leaf of a hydrophyte (*Nymphaea*) showing large intercellular spaces. Stomata are present in the upper but not in the lower epidermis.

is also an increased development of mechanical tissues. More striking are the genetically determined modifications that enable plants to adapt to particular ecological conditions, but even here the extent of expression may be directly influenced by the environment.

Hydrophytes are plants that live in the water or in extremely wet terrestrial conditions. The quantity of xylem tissue is commonly substantially reduced along with other supporting tissues, although the phloem may not be. The mesophyll is characterized by exaggerated intercellular spaces (Fig. 9.7) that help in buoyancy, the cuticle is usually very thin, and the epidermal cells often contain chloroplasts. In the case of submerged plants, stomata may be lacking or are no more than vestigial structures, and in floating leaves, such as water lily or *Nymphaea* sp. (see Fig. 9.7), they may be confined to the upper surface of the lamina exposed to the air. In some plants of an aquatic habitat, the leaves that are submerged have a distinctive form and structure different from that of leaves developed on exposed shoots. The submerged leaves tend to be larger and more dissected, with less organized internal tissues, than the aerial leaves that are exposed to the atmosphere. Experimental studies have shown that the plant hormone abscisic acid (ABA) has a role in leaf form and the associated tissue differentiation in submerged versus aerial leaves; ABA promotes the formation of aerial characteristics in submerged leaves. Thus, in plants that have two leaf types, the expression of one or the other is influenced by the environment and ABA.

Xerophytes, plants adapted to conditions of extreme aridity, are distinctly different from hydrophytes. In these plants the leaves tend to be small, sometimes to the extent that the main photosynthetic function is performed by the stem and the constituent cells are often of reduced size. In the leaves there is a high

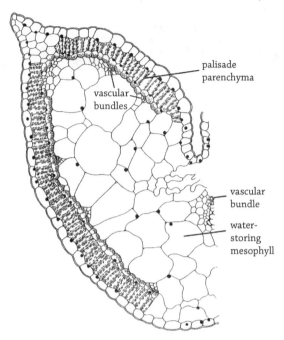

FIGURE 9.8 Diagram of a transverse section of the leaf of a halophyte (*Salsola kali*) with small vascular bundles, large water-storing cells, and palisade mesophyll on both sides of the leaf. From *Plant Anatomy*, 4th edition (1990), by A. Fahn, with permission of Elsevier Ltd.

proportion of palisade to spongy mesophyll, the walls of epidermal cells are conspicuously thickened, and the cuticle on the surface is extensively developed. The stomata are usually sunken below the level of the epidermis surface and often occur in furrows to restrict water loss (e.g., in *Ammophila* sp.; see Fig. 9.6). The vascular system is extensive and the leaf often contains a great deal of sclerenchyma. There are of course varying degrees of expression of these features and, as mentioned above, they may be exaggerated in response to drought conditions.

Finally, plants known as halophytes are adapted to conditions of high salinity. Such plants often have fleshy or succulent leaves with a high proportion of water-storing mesophyll parenchyma (Fig. 9.8). This consists of enlarged cells with large vacuoles containing cell sap or dilute mucilage. Water is easily drawn from such cells by photosynthetic palisade mesophyll, which may be present on both adaxial and abaxial surfaces (see Fig. 9.8).

LEAF DEVELOPMENT

As described in Chapter 6, leaves are initiated as outgrowths from the periphery of the SAM in a regular pattern that forms the basis for the mature phyllotaxy.

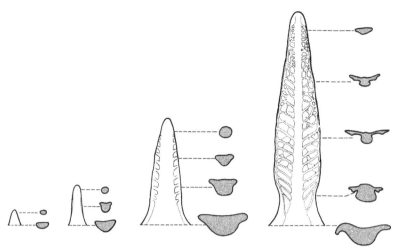

FIGURE 9.9 Early stages in the development of a tobacco (*Nicotiana tabacum*) leaf.

Accelerated cell division in several layers of the SAM results in a protrusion that, if it is large relative to the size of the apex, may be referred to as a foliar buttress. This is the beginning of a leaf primordium, which then increases in length through further cell division that may be active at, but is not restricted to, its apex. At the same time there is a progressive change in shape of the peg-like outgrowth from its original circular outline to one that is flattened on the side facing the SAM (Fig. 9.9). This causes the primordium to be bilaterally symmetrical, the initial expression of dorsiventrality that is a characteristic of most mature leaves.

The further extension of this symmetry is observed in the early initiation of the development of the lamina or blade of the simple leaf along the margins of the bilaterally symmetrical peg. These bands of meristematic activity have traditionally been called marginal meristems, but there is no clear evidence of strictly localized activity and the development of the lamina appears to be accomplished by overall growth. Very early in the process of lamina formation, distinct horizontal layers are established, and these are maintained in later leaf development by predominately anticlinal divisions in each layer (Fig. 9.10a). This process forms what is called a plate meristem and results in substantial growth of the lamina, often to many times its original extent. The layered pattern is disrupted only in localized regions by the initiation of vascular tissue and associated tissues in the venation pattern (see Fig. 9.10b). The procambium of the major veins differentiates in an acropetal and outward pattern while the smaller veins differentiate later in a basipetal pattern from the tip as the lamina develops. The procambium is apparently always in continuity. The initiation of the lamina does not extend

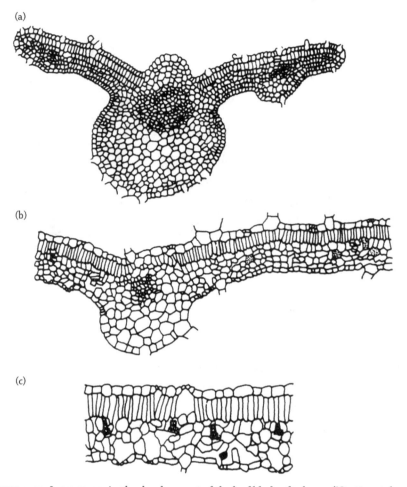

FIGURE 9.10 Later stages in the development of the leaf blade of tobacco (*Nicotiana tabacum*).

completely to the base of the primordium, leaving a zone that may later develop into a petiole.

In the case of compound leaves there is a further complication in that lamina development is not initiated in continuous strips along the margins of the peg. Rather, localized outgrowths occur on the initial peg, and these then develop in a pattern similar to that of the lamina of a simple leaf. In the case of pinnately compound leaves the sequence of initiation of leaflets may be acropetal, basipetal, or bidirectional beginning somewhere along the length of the original peg. Foliar buds often reveal, when dissected, unexpanded leaves in which the lamina is folded, rolled, or plicated in various ways. These features, which enable the primordia to be packed efficiently within the bud, are established at the time that the lamina is being initiated.

In some monocotyledons a different pattern of development leads to the formation of tubular leaves, as in onion, or unifacial leaves (because there is no distinction of adaxial and abaxial faces) as in *Iris*. In such cases, the initiation of marginal growth does not occur but there is extensive development of the abaxial face of the primordium, which results in a leaf that is radially symmetrical. The developing leaf may remain cylindrical, often becoming hollow or tubular, or may subsequently become flattened laterally to form an isobilateral leaf. The sheathing leaf base of monocot leaves is not affected by such developments.

The last phase of leaf development involves expansion of the structures already formed prior to final maturation (see Fig. 9.10c). Both cell division and cell enlargement participate in this process. Cell division continues until the leaf has attained one-half to three-quarters of its final length and then slows down as cell enlargement continues. Thus the later stage of leaf growth is mainly by cell expansion. The maturation of tissues occurs first at the tip of the lamina and proceeds basipetally. In the final phase of laminar development some of the most characteristic features of its internal structure appear as the individual parallel layers mature, but at different rates. Cell division ceases first in the two epidermal layers and then in the layers that will produce the spongy mesophyll, while it continues in the layer or layers that will become the palisade tissue. Cell enlargement ceases first in the spongy mesophyll, with the result that the cells are separated and often distorted to some extent, producing the characteristic features of this layer (i.e., large intercellular spaces). The palisade cells elongate perpendicular to the leaf surface and cease to enlarge slightly before the adjacent epidermal cells cease their growth. The result is that the palisade cells are separated from one another, albeit to a lesser degree than those of the spongy mesophyll, a feature of great significance in permitting the gas exchange necessary for photosynthesis to occur. Finally, the epidermal layers cease to enlarge. It is not uncommon for the cells of the lower epidermis to demonstrate somewhat irregular or jigsaw-puzzle shapes, reflecting the resistance to their enlargement provided by the precocious maturation of the adjacent spongy mesophyll. The precision of the control of development of these parallel layers of tissue, which is essential for the functional effectiveness of the lamina in photosynthesis, is indeed remarkable.

Thus, in contrast to the shoot, the growth of a typical leaf is determinate—that is, after a definite period of development it becomes completely mature. There are, however, instances of compound leaves, for example *Guarea glabra* (Meliaceae), in which the pinnately compound leaf continues to form leaflets from an apparently indeterminate apical meristem. Mature leaves may function for periods of several years, but in regions of seasonal climate they are typically shed or abscise after one season. The removal of senescent leaves is enhanced by a process of abscission in which the cell walls in a zone of cells, the abscission zone, located at the base of the petiole, are weakened by enzymes (e.g., cellulase) and the dissolution of middle lamella by the enzyme pectinase. The layer of cells underneath this abscission layer becomes suberized to form a protective layer that protects the exposed tissue from invasion by pathogens and water loss. Plant hormones, auxins, ABA, and ethylene have all been implicated in regulating the process of leaf abscission; auxins delay whereas ABA and ethylene enhance leaf abscission.

The development of the distinctive leaf of the grasses deserves special mention. The development of the sheathing leaf base has already been mentioned. Growth in length of the primordium is accomplished by upward growth of the margins to produce a hood-like structure of limited extent. Early in development a ligule separates the future leaf blade and the sheath. Growth in the leaf blade is by an intercalary meristem restricted to a zone of dividing cells just above the ligule, while growth of the sheath lags behind. Ultimately growth of the sheath is restricted to its base just above the node of attachment and continues for some time. Thus the result is a linear blade and an extended sheath.

Finally, the compound leaf of palms also deserves special mention because of its unique method of formation. The leaf primordium consists of a sheathing base with a hood-like precursor of a blade surmounting it. The flanking regions of the blade become folded or pleated as a result of differential growth except for the margins. After the folds become considerably deepened, separation occurs along the tops of the ridges, allowing the separated segment to become folded leaflets. The marginal strips are then shed. Thus the elaborate compound leaf is in reality a dissected simple leaf.

10

The Root

ROOTS ARE THE underground portion of the plant and, therefore, are not readily accessible for examination. This is perhaps one of the reasons that the roots have not been studied as extensively as the aerial parts. Nevertheless, roots are an essential component of the plant body, especially of terrestrial plants. They serve several important functions, primarily for the anchorage of the plant body and the absorption of water and various minerals from soil and their transport to the rest of the plant body. These functions are achieved by the enormous root system produced by plants. Indeed, in some plants the root system is more extensive than the aerial portion. Roots also serve as organs for storage of materials such as starch and sugars, as in carrot and radish. Another vital function of roots is the synthesis of certain hormones, such as cytokinins and gibberellins, which are then transported from the root to the shoot, where they effect growth and development of aerial parts. Finally, roots also have a role in vegetative or asexual reproduction of some plants (see Chapter 4). For example, in some poplar and willow species, roots produce shoot buds that grow and produce additional trees and colonize an area. Some aggressive weeds such as the leafy spurge (*Euphorbia esula*) and toadflax (*Linaria vulgaris*) spread extensively, taking over major portions of land, by producing shoot buds from roots.

ROOT SYSTEMS

The root system can be extremely simple, consisting of a main root and a few branches. However, more commonly the root system is more complex, with an extensive network of branches and with roots of different growth potential. The development of a root system depends on a number of factors, including genetic and environmental factors, the availability of moisture, and soil conditions.

In almost all seed plants, the first organ to emerge from a germinating seed is the root. This young root, the radicle, begins to grow into the soil (i.e., positively gravitropically) and is called the primary root. The extent of primary root growth varies in different plants. In many dicots and most gymnosperms, the primary root continues to grow into soil, and this growth can be as much as several meters deep. This primary root along with its many branches, called the lateral roots, forms the tap root system (Fig. 10.1a). In other dicotyledons

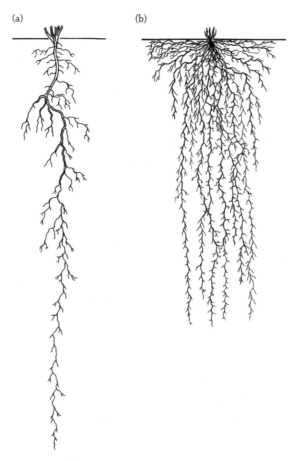

FIGURE 10.1 Root systems: (*a*) tap root system and (*b*) fibrous root system.

From *Botany: An Introduction to Plant Biology* by T. E. Weier, C. R. Stocking, and M. G. Barbour, 5th edition (1974), with permission of Dr. M. G. Barbour.

and most monocotyledons, including grasses, the growth of the primary root is limited. The root stops growing after a certain period, but it produces several branches that may be of the same length as the primary root or longer. This type of root system, which forms a localized clump, is called the fibrous root system (Fig. 10.1b). Another type of root system is formed by roots originating mainly from the stem region, nodes and internodes, immediately above the ground level. These adventitious roots help support the stem in tall grasses as in maize (*Zea mays*) and in some trees, for example the screw pines, members of the genus *Pandanus*. In some tropical plants, such as the banyan tree (*Ficus benghalensis*), the shoot branches produce aerial roots that hang down and on contact with soil form a root system. Such roots support the tree branches and have been called prop roots. In plants with trailing stems, for example the strawberry, roots are produced from nodes of the stem at regular intervals. Similarly, in certain vines, aerial roots are often produced from the stem, and they provide support to the climbing shoot.

ROOT ASSOCIATIONS
Mycorrhizae

Roots of almost all terrestrial plants have an association with soil fungi, and these associations are of mutual benefit to the plant and the fungus. The fungus absorbs minerals, especially phosphorus, from soil and passes them on to the plant, and the plant in turn provides nutrients (e.g., carbohydrates and amino acids) to the fungus. These fungal associations are called mycorrhizae (Fig. 10.2). The fungus may form a dense mat on the outer surface of the root,

FIGURE 10.2 An ectomycorrhizal root with fungus association that causes swelling of the roots in *Pinus strobus*. (Courtesy of Dr. Larry Peterson)

called ectomycorrhiza, or it may penetrate into the root and branch out, forming an association called endomycorrhiza.

Root Nodules

In some members of the pea family (Fabaceae), roots develop swellings or nodules (Fig. 10.3) caused by the invasion of a bacterium (*Rhizobium* sp.). The nodules are formed by the bacterial-induced cell division and cell expansion of cortical cells of the root. The bacteria live in the root but have a symbiotic relationship with its host; in other words, they fix atmospheric nitrogen, which plants are incapable of doing, and convert it to a useable form (e.g., ammonia) that is used in the formation of several important compounds, including amino acids and nucleotides (the building blocks of DNA and RNA), which are useful to the plant. The plant provides various stored products (e.g., carbohydrates) to the bacteria. In addition, a number of woody species, such as alder (*Alnus*) and Australian oak (*Casuarina*), form an association with filamentous bacteria, and these also fix atmospheric nitrogen for the benefit of the host.

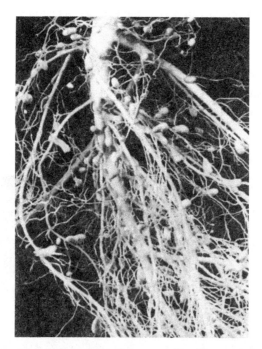

FIGURE 10.3 A legume root system with nodules caused by the invasion of a bacterium, *Rhizobium* sp.
From *Seed to Civilization* (1981) by C. B. Heiser Jr., with permission of W. H. Freeman and Co.

Haustoria

Some angiosperm species live and grow on other plants as parasites. These parasites do not develop a normal root system. Instead, they produce specialized structures called haustoria that may grow on the surfaces of shoots and roots of their host and then invade the host to draw moisture and nutrients from the plant, for example the dwarf mistletoe (*Arceuthobium* sp.). Haustoria are mycelium-like structures that penetrate the epidermis of the host plant and make connections with the xylem and, in some cases, the phloem of the parent plant. Parasitic plants are not uncommon among angiosperms.

ROOT APEX AND THE ROOT APICAL MERISTEM

The tip of the root, the root apex, is simpler in organization than that of the shoot tip. The main reason is that there are no outgrowths or organs produced directly from the root apical meristem (RAM), as leaves are from the shoot apical meristem (SAM). Thus the root apex consists mainly of the RAM. The RAM, unlike the SAM, is, however, not exposed at the tip but is protected by a covering, called the root cap. The root cap is a group of cells produced by the RAM itself and, therefore, the meristem is not terminal but sub-terminal, unlike the SAM. Because there are no outgrowths at the tip, the limits of the root apex are often difficult to define.

The structure of the RAM is varied in different groups of plants, and there are variations even within the same group. In angiosperms, the RAM commonly consists of meristematic layers, also called histogens, at the tip that form a layering pattern, but this pattern is not universal. The following types of RAMs are recognized in plants.

Single Apical Cell

In most ferns, horsetails, and species of *Selaginella*, the RAM is a single large apical cell (Fig. 10.4), which together with its derivatives produces the entire root. The apical cell is similar to that found in the shoot apex of these groups of plants (see Chapter 6); however, here the cell is like an inverted prism and has four cutting surfaces instead of three in the shoot apical cell. The lateral surfaces of the apical cell divide to produce cells that form most of the root body, including the various tissues of the root. The distal face of the apical cell divides to produce cells that form the root cap; this is in contrast to the shoot, where the distal end of the apical cell does not divide at all.

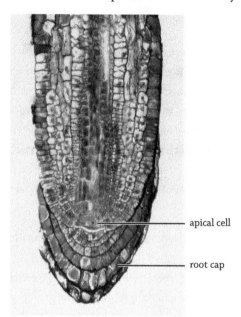

FIGURE 10.4 Longitudinal section of the root apex of the fern *Adiantum* with an apical cell that is the ultimate source of all cells and tissues of the root.

Meristem Layers

In many angiosperms and gymnosperms, the RAM commonly consists of three or four meristematic layers. In general, cells in each of these layers divide to produce a specific tissue, or tissues. In radish (*Raphanus*), for example, there are three meristem layers (Fig. 10.5). The distalmost layer (toward the tip) divides to form both the root cap and the epidermis. The second or middle layer forms the cortex, and the innermost layer produces the vascular tissue. Thus, each of the three meristem layers is destined to produce specific tissues of the root. A similar organization of the meristem is observed in other plants, including the model plant *Arabidopsis*. In corn (*Zea mays*), the RAM also comprises three layers, but the distalmost layer forms only the root cap. This special layer is also called the calyptrogen. The second layer forms both the epidermis and cortex, and the innermost layer the vascular tissue. Thus, the fate of derivatives of the three meristem layers is different in *Zea* in comparison to *Arabidopsis* and *Raphanus*. In other plants, the number of meristem layers may be two or four, although such examples are rare.

Open Meristem

In some plants, the RAM does not have distinct meristem layers, as in pea (*Pisum*) or onion (*Allium*) (Fig. 10.6). Here, the meristem consists of a group of cells at the

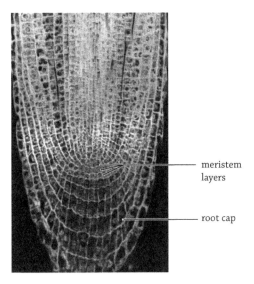

FIGURE 10.5 Longitudinal section of the root apex of radish (*Raphanus sativus*) with three distinct meristem layers: the distal layer forms the root cap and epidermis, the middle layer the cortex, and the innermost layer the vascular tissue.

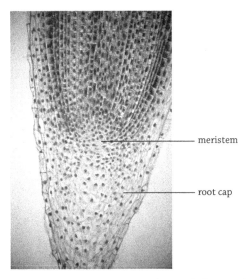

FIGURE 10.6 Longitudinal section of the root apex of onion (*Allium cepa*) with an open type of meristem (i.e., with no distinct layers).

tip of the root that divide both periclinally and anticlinally to form different tissue systems of the root. This type of RAM is found in many angiosperms and gymnosperms and is described as the open meristem.

The varied structure of the RAM in different species is further compounded in some plants where the organization of the meristem varies at different stages of development. For example, in *Helianthus*, the young root has the layering pattern of meristem whereas the older root has an open-type meristem.

One special feature of the root apex is that in the center of the RAM there is a group of cells that are mitotically less active than the surrounding cells. This region is called the quiescent center and comprises a small group of cells. Cells of the quiescent center can, however, become active under certain conditions. For example, in roots that are recovering from a period of winter dormancy or from nutrient starvation, quiescent cells become meristematic and contribute to cells of the root. Similarly, in case of an injury to the RAM, cells of the quiescent center become active in some species and divide and reform the complete meristem. Thus, quiescence of these cells is not a permanent feature, but they become mitotically active in response to changes in environmental conditions. A similar quiescent region exists in the central zone of the shoot apex of some plants (see Fig. 6.5 in Chapter 6), indicating the importance of this feature in the perpetuation of an indeterminate meristem.

TISSUES OF THE ROOT

The root, like other parts of the plant body, possesses the three tissue systems, dermal, fundamental or ground, and vascular tissues (Fig. 10.7). The basic pattern of tissue differentiation in the root is similar to that in the stem (see Fig. 8.6 in Chapter 8). That is, cells produced from the RAM differentiate into the three sub-meristems, the protoderm, the ground meristem, and the provascular tissue

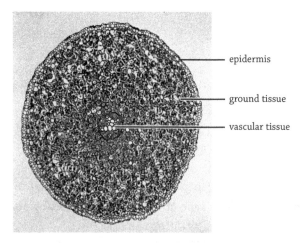

FIGURE 10.7 Transverse section of buttercup (*Ranunculus*) root with the three tissue systems: dermal (epidermis), ground, and vascular tissue.

from which the procambium is differentiated. Each of these sub-meristems then forms the three tissue systems with different cell types. Some of the cell types found in the stem are lacking in the root, for example guard cells in the epidermis and collenchyma in the cortex.

The organization of the three tissue systems varies in roots of different species, and especially between the dicotyledons and monocotyledons. In a transverse section of a root the three tissue systems are commonly recognized as follows: a single layer of dermal tissue at the periphery, inner to which is the cortex that constitutes the ground or fundamental tissue, and the vascular tissue as a cylinder or the stele in the center of the root. The xylem is located in the center of the stele and is generally in the form of a lobed column or a protostele, and the phloem is located in between the xylem arms. Although roots of most angiosperms possess a protostele, in some species, especially monocotyledons, there may be a central pith. A brief description of the arrangement of tissues and cell types commonly found in roots follows.

Dermal Tissue

The dermal tissue consists of a single layer of epidermis composed of tightly packed cells that enclose other tissues of the root. The epidermal cells have a thin layer of cuticle on the outside, which may thicken in older parts of the root. The striking feature of the epidermis is the production of root hairs, which are literally outgrowths of epidermal cells and, therefore, are unicellular (Fig. 10.8). Root hairs can be extensive and their function is to increase the surface area of the root epidermis for the absorption of water and minerals from the soil. In plants that grow underwater, root hairs are not well developed and the water is absorbed directly by the epidermis. There are no openings or pores (e.g., stomata) in the epidermis of roots such as those found in the leaves or in the stem.

In some monocots, particularly members of the orchid family and in aerial roots of some other plants, the epidermis may be more than one cell layer thick, and this multilayered epidermis is called the velamen. Cells of the velamen generally have secondary wall thickenings that prevent the loss of water as well as provide mechanical support to aerial roots.

Ground Tissue

The ground or fundamental tissue of root is the cortex and is present immediately below the epidermis. It is generally several layers thick and consists of parenchyma

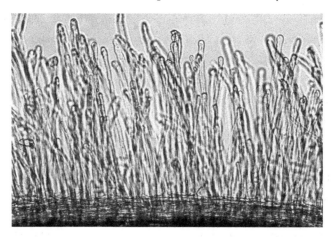

FIGURE 10.8 A portion of lettuce (*Lactuca sativa*) root with an abundance of root hairs.

cells that are devoid of chloroplasts but usually contain starch. There are intercellular spaces between cortical cells, and in plants growing underwater these spaces can be quite large and, as in the stem, the tissue is called aerenchyma. Collenchyma is rarely found in roots, although sclerenchyma may be present in some cases.

The innermost layer of the cortex is a specialized layer of cells called the endodermis (Fig. 10.9). The endodermis is a single layer of cells and surrounds the central vascular cylinder, the stele. The endodermis is an important layer as it serves the function of controlling the movement of water and nutrients from the cortex to the vascular tissue in the root. In other words, it is a kind of barrier that controls what goes into the vascular tissue of the root and then on to the aerial parts. The endodermal cells are cuboidal in shape and are tightly packed; in other words, there are no intercellular spaces between them, unlike the rest of the cortex. The unique feature of these cells is the deposition of a continuous strip of suberin, a fatty substance, in the radial and transverse walls of these cells. This band of suberin, called the Casparian strip (see Fig. 10.9b), prevents the flow of water from the cortex through the cell walls (apoplast) of endodermal cells. Thus the water, along with the dissolved nutrients, is forced across the cell membrane and cytoplasm (symplast) of endodermal cells. In this way, the endodermal cells control the permeability of substances transported to the vascular cylinder.

In some plants one or a few layers of cells immediately below the epidermis may also have suberin deposited in their walls. This is called the exodermis and has also been shown to exercise control of substances entering into the root. In older parts of the root, the epidermal cells may die, and in that case the exodermis serves as the surface layer. In the roots of some species, the cortex may develop secondary wall thickenings and thus become sclerenchyma.

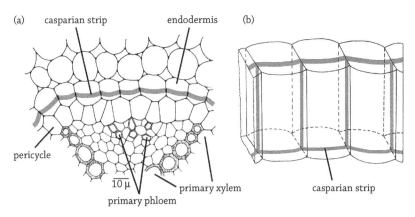

FIGURE 10.9 Diagrams of a portion of a root: (*a*) transverse view showing endodermis, pericycle, primary xylem, and primary phloem and (*b*) three-dimensional view of the endodermis showing the presence of casparian strip on transverse and radial walls.
From *Anatomy of Seed Plants* by K. Esau (1977), reprinted with permission of John Wiley & Sons Inc.

Vascular Tissue

In the root, unlike the stem, the vascular tissue is not in the form of vascular bundles or strands but, as stated above, it occupies the center of the root as a stele (see Fig. 10.7). The stele consists of the vascular tissue and a nonvascular tissue, the pericycle. The pericycle is included as part of the stele because it originates from the same meristematic layer, or the same part of the meristem, as the vascular tissue. The pericycle is located immediately inner to the endodermis and may be one or a few layers of cells (see Fig. 10.9a). Pericycle cells are commonly parenchyma cells, but in some plants they develop secondary walls and become sclerenchyma. The pericycle has several important roles; it is the layer responsible for the formation of lateral roots and, where present, shoot buds for vegetative propagation. Pericycle also contributes to the formation of the vascular cambium, and in some cases cork cambium, in plants with secondary growth.

In most dicotyledons, xylem is located in the center of the root, but it has arms or arches extending out from the center. The number of arches varies from one to several. Thus, a root may have one (monarch), two (diarch), three (triarch), four (tetrarch), five (pentarch), six (hexarch), or many (polyarch) arches (Fig. 10.10). The polyarch condition is commonly found in monocotyledons, and in these cases there is often pith present in the center of the root (Fig. 10.11). The differentiation of xylem elements in a root is centripetal—that is, the protoxylem elements are located at the tips of the xylem arms and the metaxylem elements toward the center of the root. This pattern of xylem differentiation is called exarch, in

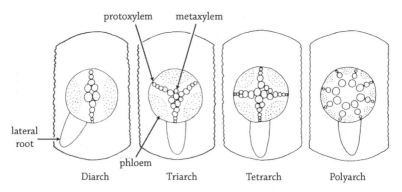

FIGURE 10.10 Diagrams of different patterns of xylem organization in the root. From *Anatomy of Seed Plants* by K. Esau (1977), reprinted with permission of John Wiley & Sons Inc.

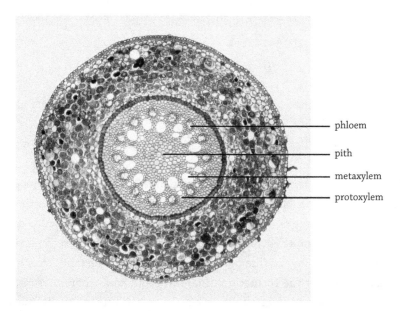

FIGURE 10.11 Transverse section of a monocot root showing xylem alternating with phloem, and with pith in the center.

contrast to endarch in the stem, and is typical of roots, although the stems of some less advanced vascular plants, for example club mosses (*Lycopodium* sp.), also have an exarch condition.

In roots, the phloem is located in positions alternating with the xylem arches so that it is present between the arches (see Fig. 10.10). In polyarch roots with pith, xylem strands alternate with those of phloem (see Fig. 10.11). The pattern of phloem differentiation is also centripetal. Thus protophloem elements are present close to the pericycle and metaphloem near the interior of the root.

Root Cap

The root cap is a special tissue present at the tip of the root; its main function is to protect the RAM (see Figs 10.4 through to 10.6). Cells of the root cap are of parenchyma and are living, and they commonly contain plastids with starch, the amyloplasts. These cells also secrete a mucilagenous substance of pectic nature that is suggested to help in the penetration of the growing root tip into the soil. As the root grows, the root cap cells are constantly lost from the tip and new ones are added by the RAM. The root cap cells, sometime called border cells, also serve as an interface between the root and fungi in the soil and are part of the rhizosphere, an area that immediately surrounds the root. Finally, the root cap has long been suggested to have a role in gravitropism, although this view is not widely held.

ROOT BRANCHING

As indicated above, the root system is largely formed by branches, which in turn may branch further, and are produced either from the primary root as lateral roots or by adventitious roots produced from the aerial parts of the plant. Root branching is extremely important for the plant because in addition to increasing the surface area for the absorption of water and minerals, it serves the major function of improved anchorage for the plant body. The origin of both the lateral and adventitious roots is endogenous (i.e., internal) in contrast to shoot branches, which are produced at the surface.

Lateral Roots

In most angiosperms and gymnosperms, lateral roots have their origin in the pericycle. A few cells in the pericycle at a certain distance from the tip are involved in the formation of a lateral root, and the pattern of origin is varied in different plants. In plants with a diarch root, lateral roots originate in between the xylem and phloem, in triarch, tetrarch, and pentarch opposite the xylem poles, and in polyarch roots opposite the phloem (see Fig. 10.10). The precise positioning of the lateral roots in some cases leads to distinct rows of roots along the primary root but, because of the twisting nature of the main root during its growth, clear rows of laterals along the primary root are generally not detected.

The initiation of a lateral root begins by divisions in a few cells of the pericycle (Fig. 10.12a). The first divisions are periclinal and are usually unequal, resulting in smaller outer and larger inner cells. These divisions are followed by anticlinal

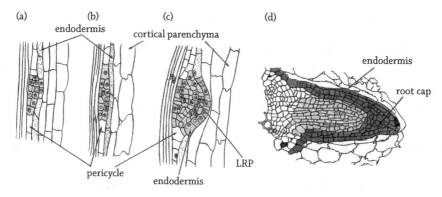

FIGURE 10.12 Different stages (*a* to *d*) in the formation of a lateral root. LRP = lateral root primordium.
From *Botany: An Introduction to Plant Biology* by T. E. Weier, C. R. Stocking, and M. G. Barbour, 5th edition (1974), with permission of Dr. M. G. Barbour.

divisions of these derivatives, resulting in a clump of cells that represent the first sign of a lateral root primordium (see Figs. 10.12b and 10.12c). The primordium begins to grow and exerts pressure on the overarching endodermis, which in turn divides, mostly by anticlinal divisions, to keep pace with the enlarging root primordium. With further divisions, the developing primordium begins to push its way through the cortex, crushing the cortical cells and the epidermis (see Fig. 10.12d), and ultimately emerges at the surface of the root. During the development of a lateral root, cells of the xylem and phloem parenchyma of the parent root divide and contribute to the formation of vascular tissue of the lateral root, thereby establishing a vascular connection with them. Also, before the lateral root emerges on the surface, the RAM is well established and with a quiescent center and root cap.

Adventitious Roots

Roots that originate from parts of the plant other than the root (i.e., stem, leaf blade, petiole, or peduncle) are called adventitious roots. Adventitious roots are found in many monocotyledons, in some dicotyledons with underground stems, runners, and aerial vines, and in plants grown in water. The origin of adventitious roots is also endogenous. In a stem, they commonly originate from the interfascicular parenchyma, which is the region between the vascular bundles. In a woody stem, adventitious roots originate from ray parenchyma (see Chapter 11). The early stages in the formation of adventitious roots are similar to those of lateral roots, although, as stated before, the tissues involved are different.

The ability to induce adventitious root formation artificially is of importance in the areas of horticulture and forestry as it serves a useful method for vegetative propagation of elite plant varieties. Auxins, both natural and synthetic, have been shown to stimulate the initiation of both adventitious and lateral roots and are commonly used for rooting of cuttings, for example of many fruit trees and ornamental plants. However, auxin itself inhibits the growth of the root primordia formed and needs to be washed from the stem or other tissues for promoting the growth of secondary roots. This also indicates that the formation of lateral root primordia and their subsequent growth are two separate processes requiring different controlling mechanisms.

SHOOT BUDS FROM ROOTS

In many plants, roots also produce shoot buds (Fig. 10.13), which grow and form fully developed shoots and mature plants as a means of vegetative reproduction (see Chapter 4). Shoot buds may be induced in response to injury to the root, but in other cases this is a natural phenomenon, and in many plants that grow in the subtemperate or temperate regions, this is the chief method of reproduction. For

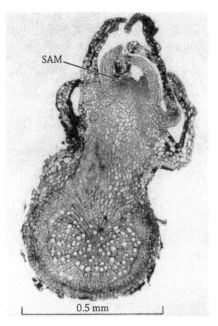

FIGURE 10.13 Shoot bud regeneration from the root of *Viola adunca*. The root is shown in cross-section (lower part) and a shoot with apical meristem (SAM) and leaf primordia is present in the upper part.

From Raju et al. (1966), *Canadian Journal of Botany* 44: 33, with permission of the NRC Research Press.

example, the trembling aspen (*Populus tremuloides*), which grows on the prairies or grasslands, reproduces mainly by this method, whereas in the boreal forest it reproduces sexually by the formation of seeds.

The origin of shoot buds is also endogenous like that of lateral roots and, interestingly, from the same tissue as the laterals. A few cells in the pericycle, usually near the protoxylem, begin to enlarge and divide periclinally. The initial periclinal divisions result in three or four layers of cells in which later anticlinal divisions occur. These layers of cells initially form a SAM, and later leaf primordia are initiated at the margin of the SAM resembling a shoot apex (see Fig. 10.13). Unlike the lateral root, the endodermis does not participate in the formation of a shoot bud. The developing young bud pushes its way through the root cortex and appears at the surface of the root with an intact meristem and a few leaf primordia. Initially the shoot bud establishes vascular connections with the parent root, but soon it develops its own root system.

The question of what determines whether the pericycle cells will form a root or a shoot is not completely understood. Experimental work with isolated root segments indicates that there is a polarity in the root axis; the distal end of the root (toward the root tip) regenerates only roots, whereas the opposite proximal end initiates shoots. This polarity may have its basis in the relative concentration of auxins to cytokinins in different parts of the root; a lower ratio of auxin to cytokinin toward the proximal end favors shoot formation, whereas a higher ratio toward the distal end promotes roots.

11

The Secondary Body

THE CELLS AND tissues produced by the apical meristems of shoot and root, including everything that has been discussed thus far, constitute what is designated the primary body of the plant. In many instances this comprises the entire plant, but in other cases, particularly plants with an extended lifespan, there may be two additional meristems, the vascular cambium and the cork cambium or phellogen, which produce the secondary body. Unlike the apical meristems, these meristems are located in a lateral position in the stem, the root, and occasionally the leaf petiole and produce only selected tissues that supplement those of the primary body. The vascular cambium initiates secondary vascular tissue, xylem and phloem, and the phellogen gives rise to the secondary dermal system or periderm. As was pointed out in Chapter 1, plants, because of their cell structure, cannot replace worn-out cells like animals; instead they provide a continuing supply of new functional cells that are especially important for plants with a long lifespan. Whereas the apical meristems increase the height or stature of plants, the secondary meristems increase the width or diameter and in large trees and shrubs may actually contribute the bulk of the tissue present in older individuals. However, secondary growth is also often present in smaller herbaceous plants, even in many annuals. It is found in all gymnosperms and in a large number of dicotyledons but is absent in most monocotyledons.

THE VASCULAR CAMBIUM
Initiation of the Vascular Cambium

The differentiation of primary vascular tissue in the stem was described in Chapter 8. If a procambial strand is followed basipetally (i.e., toward the base), it may be observed that final differentiation of tissues does not occur uniformly across the bundle. Xylem differentiation is first noted adjacent to the pith, the endarch pattern, whereas the first phloem differentiates at the periphery adjacent to the cortex. The further differentiation of xylem is centrifugal, from inside to outside, and that of the phloem is centripetal, form outside to inside—in other words, differentiation of both tissues proceeds toward the center of the bundle. While this differentiation is proceeding, longitudinal cell divisions continue in the bundle, increasing the amount of procambium. Progressively these divisions become oriented parallel to the stem surface—that is, periclinally. If mature xylem and phloem meet in the interior of the bundle (i.e., if all of the procambium differentiates), there will be no vascular cambium formed and thus no secondary growth. If, however, the two waves of differentiation of xylem and phloem do not meet but rather leave a band of dividing cells between them (Fig. 11.1, arrow), then secondary growth is initiated. This activity may remain restricted to individual bundles, thus producing a limited quantity of secondary vascular tissue, but more commonly cell division activity spreads from the dividing zones in the bundles to the regions between bundles (i.e., interfascicular parenchyma), forming a continuous ring.

As noted above, the continuing divisions in the procambium become increasingly periclinal in orientation as the dividing zone is restricted by advancing

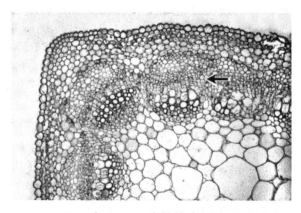

FIGURE 11.1 Transverse section of a portion of alfalfa (*Medicago sativa*) stem showing the beginning of cambial activity (arrow) in between the differentiating xylem and phloem elements.

differentiation. Ultimately distinct rows of cells can be followed from the xylem across the dividing zone into the phloem so that the zone is producing both tissues. In fact it has been concluded that in each row consisting of xylem, dividing cells, and phloem there must be a single initial cell that by its periclinal divisions functions as the ultimate source of both tissues in that row; however, the derivatives of this cell also divide to produce additional cells. All these cells, which function in this role but are usually very difficult to identify, are designated cambial initials, while the zone consisting of these initials and their dividing derivatives is designated the cambium. Support for this interpretation is found in certain plants with seasonal growth in which, during the dormant season, the cambium is reduced to a single band of cells.

The initiation of the cambium in roots presents a somewhat different picture because of the configuration of the stele. In roots, there is frequently a star-shaped central xylem core that differentiates first at the tips of the radiating arms while the phloem is located in the bays between the xylem arches and differentiates toward the center (Chapter 10). The cambium originates in the procambial region on the same radii as the phloem and lateral to the xylem arms and extends around the protoxylem poles by extension of divisions into the pericycle.

Organization of the Vascular Cambium

In examining structural features of the primary body sections are cut either transversely or longitudinally. Because of the organization of the secondary body two specific planes of longitudinal sections must be distinguished, radial and tangential (Fig. 11.2). Radial sections are cut along a radius of the stem or root and tangential sections lie at right angles to a radius. Sections cut in other planes are difficult if not impossible to interpret and result in confusion.

Little is revealed of the structure of the cambium in either transverse or radial sections other than to demonstrate the derivation of conducting, supporting, and storage elements from it. It is in tangential sections, which expose essentially a face view of the cambium, that its organization is best understood. In tangential sections two distinct cell types are evident (Fig. 11.3). Fusiform initials are elongated cells with tapering and overlapping ends that, in conifers and in some dicotyledons, may be several millimeters in length, but considerably shorter in most dicotyledons, especially advanced dicotyledons. These cells have relatively large nuclei and, surprisingly, are often highly vacuolated. Derivatives of fusiform initials differentiate into xylem and phloem, the conducting and supporting elements, as well as vertical rows of parenchyma cells. Interspersed among the fusiform initials are the ray initials, which occur in uniseriate vertical rows

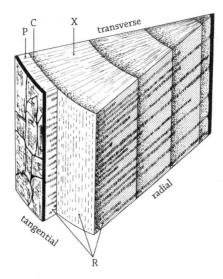

FIGURE 11.2 Three-dimensional diagram of a portion of the stem of *Pinus* exposing the transverse, radial, and tangential planes of the stem. C = cambium, P = phloem, R = rays, X = xylem.

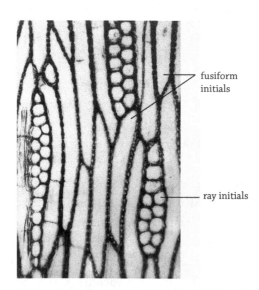

FIGURE 11.3 Tangential view of ash (*Fraxinus americana*) cambium with fusiform and ray initials. Note primary pit fields on radial walls of fusiform initials.

in conifers and in both uniseriate and multiseriate groups in dicotyledons (see Fig. 11.3). These give rise to horizontal bands of parenchyma cells that extend radially from the cambium through both the xylem and the phloem. Thus this meristem, the vascular cambium, is organized in a very different way from the

The Secondary Body

apical meristems that give rise to the shoot and root in the primary body in that its cellular components are specialized in relation to the derivatives produced.

Cambial Activity

The predominant plane of division in the cambium is periclinal, giving rise to derivatives that differentiate as mature elements of the secondary xylem toward the inside and phloem toward the outside. This activity necessitates that the cambium increase in circumference as the core of secondary xylem continues to enlarge, particularly in long-lived woody plants. A comparison of young and old stems also demonstrates that the number of fusiform initials increases with age, as does the number and width of ray initial groupings. This increase in initials must occur by anticlinal divisions adding to the number of cells. In some woody dicotyledons, such as black locust and persimmon in which the fusiform initials are very short, examination of the cambium itself reveals how the increase is accomplished because the initials are arranged in horizontal tiers reflecting repeated, longitudinally oriented anticlinal divisions (Fig. 11.4). This is called storied or stratified cambium. In most species, however, no such pattern is found, but because the secondary xylem is ordinarily well preserved it

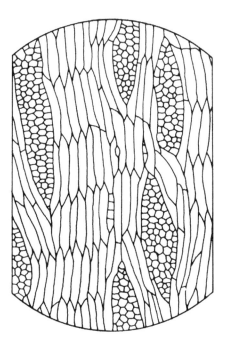

FIGURE 11.4 Tangential section of locust (*Robinia pseudoacaia*) cambium with the storied arrangement of fusiform initials.

is possible to resolve this by examining successive increments of this tissue. In the elongated overlapping (non-storied) fusiform initials anticlinal divisions also occur but they are not longitudinal in orientation. Instead they show varying degrees of obliquity ranging from nearly transverse to essentially vertical as in storied cambium. These divisions reduce the length of the fusiform initials, which subsequently elongate by a growth process in which the pointed tips grow between the walls of other initials until the original length is restored. Anticlinal divisions in the ray initials increase the width of the rays, and new groups of ray initials are also derived by divisions from fusiform initials. Thus the vascular cambium is clearly a changing and dynamic meristem that maintains its organization while undergoing extensive size increase over an extended time period.

In regions where periods or seasons favorable to growth alternate with unfavorable interludes the activity of the cambium may be periodic—that is, it may cease to function during an unfavorable period and resume its activity when suitable conditions return. Very often environmental conditions differ at the end of an active period and the beginning of the next as, for example, in late summer or autumn compared to early spring of the next cycle. The result of this is that the secondary xylem differentiated during the two periods shows recognizable differences in such features as cell diameter, wall thickness, and relative abundance of different elements. Since the transition from active growth to dormancy is gradual while that from inactivity to active growth is usually abrupt, distinct rings are formed, each one representing one season (Fig. 11.5a). These zones or rings are referred to as growth rings and, if the seasonal cycle occurs only once each year, as annual rings. These rings may then be used to determine the age of a tree and,

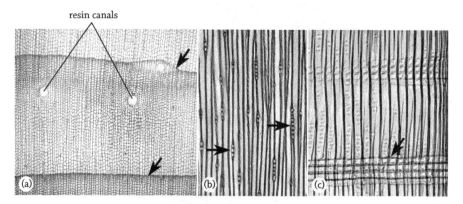

FIGURE 11.5 Secondary xylem of *Pinus*. (*a*) Transverse section showing growth rings (arrows) and resin canals. (*b*) Tangential section showing tracheids and rays (arrows). (*c*) Radial section showing tracheids with border pits along the radial walls, and rays (arrow).

since the thickness of a ring depends upon growth conditions during the period of its formation, it may also provide evidence as to past climatic conditions.

SECONDARY XYLEM (THE WOOD)

Although the cell types that compose the secondary vascular tissues are essentially the same as the components of the primary system, the organization is substantially different and reflects the structure of the meristem from which they are derived. The cell types may be thought of as two intersecting systems, one vertical, derived from the fusiform initials, and the other horizontal, derived from the ray initials. It is also necessary to examine transverse sections and both radial and tangential longitudinal sections of these tissues for a complete interpretation. In the descriptions that follow particular attention will be paid to conifers and woody dicotyledons, the two groups in which secondary growth is best developed. It is also in these two groups that the most extensive investigations have been carried out, largely because of the economic importance of the wood derived from them.

The continued activity of the vascular cambium over many years in long-lived species has the result of building up a large volume of secondary xylem. There is good evidence that only the more recent increments are actually active in the transport of water; the remainder simply supplies mechanical support for the ever-enlarging plant body. The active outer wood is referred to as sapwood while that in the interior is known as heartwood. In many cases the heartwood seems to become a repository for waste products, and these are often highly colored. It is also believed that these substances may inhibit the growth of microorganisms that would destroy the tissue and weaken the tree as a whole. Heartwood is frequently highly valued for special purposes such as the manufacture of furniture because of its characteristics.

In the conifers the secondary xylem is relatively simple in construction, consisting of tracheids derived from fusiform initials and uniseriate rays initiated by the uniseriate groups of ray initials in the cambium (see Fig. 11.5 b, c). In some species derivatives of the fusiform initials may become transversely septated and differentiate into strands of parenchyma in the vertical system. Many conifers also have resin canals lined with secretory parenchyma in the vertical system as well as in certain rays (see Fig. 11.5a).

The secondary xylem of dicotyledons is considerably more complex than that of conifers and also highly diverse, features that are useful in classification and in the identification of unknown samples. The easily traced radial rows of elements

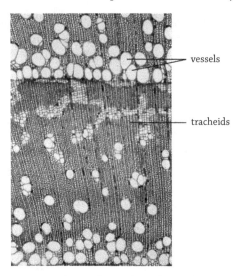

FIGURE 11.6 Transverse section of secondary xylem of *Rhus* shows ring porous wood with tracheids and vessels. Note some vessels are single and others are grouped.

derived from the cambium seen in conifers are seldom seen here because of the heterogeneity of tracheary elements—in other words, tracheids and vessels (Fig. 11.6). In all but a few species, which are believed to be primitive, vessels are present as the main conducting structures in dicotyledons. Vessel elements have been shown to represent an evolutionary derivation of the more primitive tracheid. Vessels, which are composed of superimposed vessel elements separated by perforation plates to the free flow of water (see Chapter 7), vary in length from a few feet to the entire extent of a tree several feet tall. The perforation plates range from highly oblique to transverse in orientation and may or may not be interrupted by bars of secondary wall. These features are best observed in radial sections. As viewed in transverse sections, vessels may occur singly but more often occur in various groupings, either horizontal bands or a cluster of various sizes (see Fig. 11.6). In species with growth rings reflecting seasonal activity of the cambium, the first part of the increment may be characterized by increased size or number of vessels or both and later in the season smaller size of vessels (see Fig. 11.6). This condition with alternating large and small pore vessels is described as ring porous. Alternatively, where vessel size and number are uniform through the growth rings or show only a gradual change across each ring, the wood is said to be diffuse porous (Fig. 11.7).

In addition to vessels, many dicotyledons have retained tracheids in their secondary xylem as supplementary channels of water transport or perhaps as a water-storage mechanism. However, more common are other derivatives of

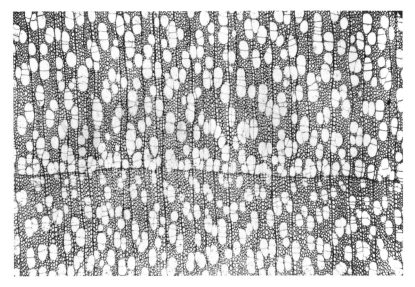

FIGURE 11.7 Transverse sections of secondary xylem of *Populus* with diffuse porous wood.

tracheids that are specialized as mechanical supporting elements. These are fibers and are characterized by greatly thickened secondary walls and a corresponding reduction in the cell lumen and the size, frequency, and functional capacity of pits. A series of these imperforate elements is recognized ranging from tracheids to fiber tracheids to libriform fibers based on wall thickness and pit development.

Parenchyma derived from fusiform initials is a very common but not a universal feature of the vertical system of secondary xylem, where it serves a storage function. Its abundance varies greatly from a very sparse distribution to an overwhelming dominance in such cases as the lightweight balsa (*Ochroma* sp.) wood. The distribution of parenchyma as observed in transverse sections may have no consistent relationships to the distribution of vessels and may be scattered, arranged in tangential bands, or, if growth rings are present, located at the end of each ring. Alternatively, parenchyma may be associated with vessels either in tangential strips that regularly include them or in clusters surrounding the vessels. The latter distribution is thought to reflect the need of the living parenchyma cells for a supply of water in wood in which there is an abundance of nonconducting fibers.

The horizontal system of the secondary xylem is usually represented by rays of two types, although there are cases in which only one type is present. Multiseriate rays may be relatively low in height and only a few cells wide (Fig. 11.8), or in some cases they may be extremely large both vertically and in width. Uniseriate rays are only one cell in width, as in conifers, but vary greatly in vertical height.

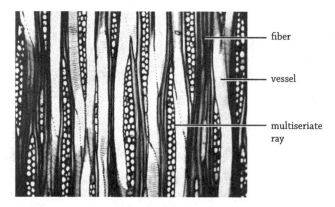

FIGURE 11.8 Tangential section of *Fraxinus americana* wood with multiseriate and some uniseriate rays.

Studies in which the pattern of rays has been followed through successive increments of secondary xylem have revealed significant changes with time. When secondary growth begins at any level, typically multiseriate rays arise opposite the interfascicular parenchyma between the vascular bundles of the eustele whereas uniseriate rays develop from procambium in the bundles. As subsequent growth progresses, additional uniseriate rays may arise from fusiform initials. Other changes in rays include the following: uniseriate ray initials may be converted to multiseriate by anticlinal divisions, existing multiseriate initials may enlarge by cell divisions, and multiseriate rays may be divided to form additional rays by the intrusion of elongating fusiform initials. These events contribute to the increase in circumference of the cambium around the growing internal mass of secondary xylem along with the increase in fusiform initials previously described.

SECONDARY PHLOEM

Secondary phloem is initiated by the same meristem that forms the secondary xylem (i.e., the vascular cambium); thus it is not surprising that there are many similarities in organization between the two tissues. Both have vertical and horizontal systems arising respectively from the fusiform initials and the ray initials. But whereas the secondary xylem to the interior of the cambium is preserved and continues to accumulate, the secondary phloem is progressively pushed to the exterior, and although it may be extensive (Fig. 11.9), it is ordinarily distorted by the stress to which it is subjected and not preserved for an extended period. The secondary phloem of conifers, like the secondary xylem, is relatively simple in organization. The conducting elements are elongated, overlapping sieve

The Secondary Body

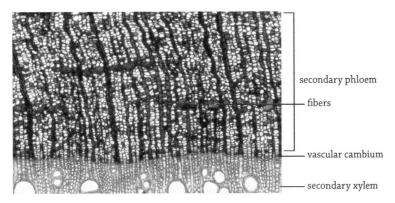

FIGURE 11.9 Transverse section of *Fraxinus americana* stem showing extensive secondary phloem.

cells that are ordinarily arranged in radial rows. These are interspersed with parenchyma cells that may be scattered or in some cases organized in tangential bands. Also occurring in tangential bands in many cases are elongated and thick-walled fibers, but not all conifers contain these. The horizontal system consists of uniseriate rays that arise from the same groups of ray initials in the cambium that produce the xylem rays. As in the xylem, resin canals may be present in the vertical system or in both the vertical and the horizontal systems. If present in the horizontal system they are enclosed in enlarged rays, the only rays that are multiseriate.

In the dicotyledons the secondary phloem is somewhat more complex and more diverse than that of conifers. As in the primary phloem, the conducting cells are sieve tube elements organized in vertical series to form sieve tubes (see Chapter 7). Although, unlike vessel elements, they are living cells, they lack a nucleus and the protoplast is highly modified. Each element is accompanied by one or more companion cells that have a common origin with the sieve element and are believed to facilitate the entry and removal of organic materials to and from it. The elements in a sieve tube are separated by sieve plates distinguished by the fact that the sieve areas are more developed than those on the side walls. Sieve plates may be transverse or variously oblique in orientation and in the latter case are inclined to face radially. Thus radial sections are ordinarily examined to determine the nature of the sieve plates. Sieve plates may have only one large sieve area or several, especially in the case of elongated, oblique plates. The vertical system also contains parenchyma that largely performs a storage function, and this may be scattered or gathered in tangential or radial bands. Crystals are often found in the parenchyma cells. Fibers are often but not always present, and these too may be scattered (see Fig. 11.9) or in tangential or radial bands.

Sieve tubes may form tangential bands alternating with bands of parenchyma and fibers may occur in long radial rows. In the radial system both uniseriate and multiseriate rays are typically present, corresponding to those in the secondary xylem, but one or the other may be lacking. In some dicotyledons, notably *Tilia* (linden), certain rays undergo considerable widening or dilation outside the cambium as a result of both cell division and enlargement, separating the remaining phloem into wedge-like segments. It is thought that this ray dilation enables the phloem to remain functional for an extended period since it relieves the strain caused by the internal expansion of the xylem core.

SECONDARY GROWTH IN MONOCOTYLEDONS

Most monocotyledons lack secondary growth of the type that occurs in conifers and dicotyledons. They do, however, have a mechanism for increasing their girth that permits some plants like palms to achieve a substantial stature. This is regarded as a continuation of primary development and is called the primary thickening meristem. It consists of a zone of periclinally dividing cells extending basipetally from the SAM for varying distances, depending upon the species. Outer derivatives of this zone differentiate as parenchyma and those on the interior form discrete bundles of xylem and phloem embedded in parenchyma. The activity of this meristem may be limited or it may lead to a substantial thickening of the axis in the region below the SAM. Furthermore, in some cases cell division and enlargement among the derivatives of the meristem may continue for some time and at some distance from the apex.

In some woody monocotyledons such as *Yucca, Agave*, and other members of the lily group, there may be a further development that much more resembles secondary growth in dicotyledons but is not homologous to it. This is called the secondary thickening meristem, which occurs outside the primary vascular system and functions much as does the primary thickening meristem except that it extends further down the stem and is long-lasting in function rather than transitory. It produces derivatives both to the outside and the interior, as does the primary thickening meristem, and these differentiate in much the same way: parenchyma to the exterior and vascular bundles embedded in parenchyma on the inner side. In some cases this secondary meristem is actually continuous with the primary thickening meristem, but in other cases it is initiated separately and at a lower level. Whether or not it should be considered a distinct secondary meristem is debatable, as it clearly is very different from the vascular cambium of dicotyledons.

PERIDERM

Where there is any significant activity of vascular cambium forming an expanding core of secondary vascular tissue, the epidermis sooner or later is ruptured, thus exposing the inner tissues it protects. The protective function is then assumed by the products of another secondary meristem, the cork cambium or phellogen, which produces a tissue known as cork or phellem to the outside (Fig. 11.10). The cork cambium commonly arises at any level along the axis, after the establishment of vascular cambium at that level, by periclinal divisions in cells immediately beneath the epidermis. However, these initiating divisions may occur in the epidermis itself or in deeper layers in the cortex. The initiation may occur in a continuous layer around the axis or it may begin at separate loci that later join. After the first periclinal division typically the inner cell differentiates to form a parenchymatous cell called phelloderm, while the outer cell becomes a cork cambium initial. In later divisions of the cork cambium initial the inner cell remains meristematic, while the outer differentiates to form a cork cell (see Chapter 7). There may or may not be subsequent additions to the phelloderm. Thus, there is generally much more cork produced than phelloderm in the stem. The phelloderm, cork cambium, and the cork together constitute the periderm (see Fig. 11.10). In roots, cork cambium initiates by a similar process occurring in subepidermal or cortical cells or even in the pericycle.

The original cork cambium may be retained for an extended period, as in oak, where it is apparently permanent. In such cases anticlinal divisions in this meristem increase its circumference. In most cases, however, it is rather quickly replaced by new cork cambia that arise inside the original cork cambium, originating in the secondary phloem from rays and parenchyma in the vertical system. As this process continues the tissue outside the active cork cambium is not only composed of cork but also incorporates everything cut off by the inward-moving

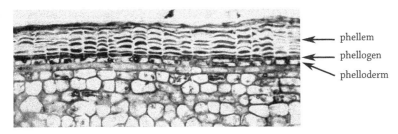

FIGURE 11.10 Transverse section of linden (*Tilia americana*) stem showing the cork cambium (phellogen) and its derivatives phellem (cork) toward the outside and phelloderm to the inside.

cork cambia. This complex mixture is termed outer bark or rhytidome, a name derived from the Greek word for wrinkle. The successive cork cambia may arise in localized, discontinuous, but overlapping arcs, with the result that the outer bark tends to separate in scale-like plates, a pattern known as scale bark. Alternatively the successive cork cambia may arise as continuous layers around the axis, producing a condition known as ring bark, as in birch (*Betula*).

The formation of impervious periderm poses a problem for the plant because the inner tissues require aeration in order to carry on their functions. This is achieved by the formation in the periderm of structures known as lenticels because of their often lens-shaped form (see Fig. 11.11a). These are formed by the cork cambium, which in localized areas produces masses of loose aerenchyma called complementary cells to the outside (see Fig. 11.11b). The lenticels that are formed in the initial periderm at any level often occur in relation to individual stomata or groups of them. Later-formed lenticels often seem to confront the outer ends of rays, particularly large multiseriate rays in the secondary vascular tissue, which would allow the passage of gases to the interior tissues through intercellular spaces in the rays. It is thought that the occurrence of ray dilation, the widening of the outer portions of rays, may enhance this interaction with lenticels. Unlike stomata, which permit the entrance and exit of gases in the primary body, lenticels have no mechanical opening or closing mechanism. Nevertheless, under conditions of stress or dormancy they may be sealed by the production of normal cork by the cork cambium and later forced open by the production of complementary cells.

In many monocotyledons there is no specific protective tissue other than the original epidermis. However, if this layer is damaged underlying cortical cells may

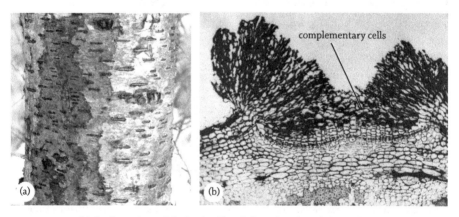

FIGURE 11.11 (*a*) Surface view of the bark of birch (*Betula* sp.) with lens-shaped lenticels. (*b*) A section through a lenticel filled with complementary cells. (Fig. *a* Courtesy of Dennis Dyck)

become suberized, thus providing the protective function. Even in the absence of damage, epidermal and some underlying cells may become suberized or lignified in long-lived organs. The monocotyledons that undergo apparent secondary growth initiate a periderm that in formation and structure resembles that of woody dicotyledons. Others that attain substantial size without secondary growth, such as palms, form a periderm in a somewhat different manner. Certain cells in the cortical region divide periclinally, forming short rows of cells that do not enlarge. These cell rows and the intervening undivided parenchyma cells then become suberized. Thus there is no recognized phellogen layer in monocotyledons with secondary growth.

GLOSSARY

ABAXIAL SURFACE The surface of a structure facing away from the shoot tip.

ABSCISSION ZONE A zone of three or four layers of cells at the base of a leaf petiole that comprise thin-walled cells; the region where cells break down during leaf abscission.

ADAXIAL SURFACE The surface of a structure facing toward the shoot tip.

ADVENTITIOUS ROOT Root produced from structures other than a root (e.g., stem, leaf or petiole).

AERENCHYMA Parenchyma cells with relatively large intercellular spaces. Useful for internal aeration of the tissue and for flotation of a structure.

AMYLOPLAST A starch containing plastid.

ANATOMY The study that deals with the internal structure of an organism or an organ and includes the organization of cells and tissues.

ANDROECIUM All the stamens in a flower.

ANNUAL RING See *Growth ring*.

ANNUALS Plants that grow for one season.

ANTICLINAL CELL DIVISION Cell division at right angle to the surface of a structure.

ANTIPODAL CELLS Cells in an embryo sac of angiosperms that are located at the chalazal end.

APICAL DOMINANCE Control of lateral bud growth in a shoot by the shoot apex.

APOCARPOUS When all the carpels in a flower are separate (free carpels).

APOMIXIS Formation of a viable embryo and seed in the absence of meiosis and without the act of fertilization.

APOPLAST The non-protoplasmic continuum between cells; it includes the cell wall, middle lamella, and intercellular spaces between cells.

ATACTOSTELE A type of stele with many scattered bundles related to leaf traces, smaller ones to the periphery and larger ones toward the inside of a stem. Commonly found in the stem of monocotyledons.

AWN An elongated outgrowth from the lemma in a grass flower.

AXILLARY (OR DETACHED) MERISTEM Meristem located in the axil of a leaf that gives rise to a branch in a shoot.

BERRY A type of fleshy fruit that develops from a superior ovary (e.g., tomato).

BICOLLATERAL VASCULAR BUNDLE A vascular bundle with phloem toward both the outside and inside of xylem; occurs in some stems.

BIENNIALS Plants that grow for two seasons.

BISPORIC EMBRYO SAC When two potential megaspores participate in the formation of an embryo sac.

BORDERED PIT Formed by the overarching of secondary cell wall over the pit membrane, leaving a small opening. Found in tracheids of some gymnosperms.

BRACT A leaf-like structure below some flowers.

BUNDLE SHEATH One or two rings of generally parenchyma cells surrounding a small vein in leaf blade.

CALLOSE A carbohydrate made up of β, 1-3 glucose molecules; commonly found in the sieve plate.

CALYPTROGEN The distalmost layer in the root apical meristem of some monocotyledons that forms only the root cap.

CAPITULUM A type of inflorescence in which several flowers are produced on an enlarged apex. Also called *head* (e.g., sunflower).

CARPEL Female reproductive organ in an angiosperm flower that consists of a stigma, style, and ovary.

CARYOPSIS A type of fruit that is nondehiscent and has a single seed, and its pericarp is fused to seed coat (e.g., in wheat).

CASPARIAN STRIP A strip of suberin, a fatty substance, in the radial and transverse walls of endodermis cells.

CATAPHYLL (BUD SCALE) Outer modified leaf of a shoot bud that has a role in protection of the bud.

CENTRAL CELL The large cell of an angiosperm embryo sac that contains the polar nuclei and encloses the egg apparatus and antipodal cells.

CENTRAL ZONE A group of cells in the center of shoot apical meristem in which cell divisions occur less frequently (i.e., at a slower rate) than in surrounding cells of the peripheral zone.

CLADOPHYLL Stem with a leaf-like form (i.e., it is flattened and broad) that performs the photosynthetic function (e.g., in some cacti).

COLEOPTILE A thin layer of tissue that encloses the epicotyl axis in a monocotyledon embryo.

COLLATERAL VASCULAR BUNDLE A vascular bundle with phloem toward the outside and xylem on the inside of a stem.

COLLENCHYMA Cell type or a tissue with primary cell wall but the wall is unevenly thickened, usually at the corners or where there are intercellular spaces.

COMPANION CELL A sister cell to sieve tube element, is smaller in size to it, has a full complement of organelles, including nucleus, and is believed to help in the functioning of sieve tube element.

COMPOUND LEAF A leaf blade (lamina) that is composed of several units (leaflets).

Glossary

CORK See *Phellem*.
CORK CAMBIUM (PHELLOGEN) A meristem that produces the secondary dermal tissue—the periderm—in plants with secondary growth.
CORPUS In the angiosperm shoot apical meristem, a group of cells below the outer tunica layer(s) in which cell divisions are in no preferred plane.
CORYMB A type of inflorescence in which flowers have stalks of unequal length but the flowers are extended to the same level.
COTYLEDON A leaf-like organ in plant embryos.
CUTICLE A layer of cutin, a fatty substance, deposited on the outside of epidermal cells. It helps in the protection of, and loss of water from, the epidermis.
DERMAL TISSUE One or more layers of the outermost tissue—the epidermis, which has several roles, depending on the structure, including protection, gas exchange, absorption of water and nutrients, and secretion of substances.
DIARCH Primary xylem in a root that has two arches, each with its protoxylem.
DICTYOSTELE A type of stele that is broken up by large leaf gaps and the vascular strands are not associated with leaf traces. Found in the stem of many ferns.
DIFFERENTIATION The process, or processes, by which cells and tissues become structurally and functionally different.
DIFFUSE POROUS WOOD The wood in which the vessel size is similar and vessels are relatively uniformly distributed.
DIOECIOUS A condition where there are separate male and female plants in a species (e.g., *Cannabis*).
DOUBLE FERTILIZATION In angiosperms, both the sperms from a pollen grain are used during fertilization; one fuses with the egg and the other with polar nuclei.
DRUPE A type of fleshy fruit with a stony endocarp around a single seed (e.g., peach).
EGG APPARATUS Consists of an egg and two synergids in an embryo sac of angiosperms.
ELAIOPLAST An oil-storing plastid.
EMBRYO SAC Female gametophyte located in an ovule in an angiosperm flower.
ENDARCH A condition when the first maturing elements (protoxlem) are toward the center and later-maturing ones (metaxylem) are toward the outside; pattern commonly found in a stem.
ENDOCARP The inner part of fruit wall (pericarp).
ENDODERMIS The innermost layer of cortex in the root, and stem of some plants, that controls the entry of substances into the vascular tissue.
ENDOSPERM A nutritive tissue in an angiosperm seed that is commonly 3n but can be 4 or 5n. It is produced by the fertilization of a sperm and polar nuclei in an embryo sac.
ENDOTHECIUM A layer of tissue in the anther in which cells have thickened radial walls; has a role in the opening of an anther.
EPICOTYL Shoot axis above the point of cotyledon attachment.
EPIGYNOUS Flower with sepals, petals, and stamens attached above the ovary—in other words, the ovary is located below them (inferior). The lower parts of floral organs, other than ovary, are fused to form a floral tube (hypanthium).
ERGASTIC SUBSTANCES Non-protoplasmic components of a protoplast (e.g., starch, lipids, tannins, and crystals).
EUSTELE A type of stele with a ring of vascular bundles (strands) related to leaf traces, commonly found in a dicotyledon stem.

EXARCH A condition when the first maturing elements (protoxylem) are toward the outside and the later-maturing ones (metaxylem) are toward the inside; pattern commonly found in a root.

EXINE The outermost wall of a pollen grain, which is variously sculptured in different species. It contains sporopollenin, a complex polymer, which makes it a tough wall resistant to decay and helps in pollen preservation.

EXOCARP The outer part of the fruit wall (pericarp).

EXODERMIS The layer of cells immediately below the epidermis in a root that has suberin in its cell walls.

EXPANSINS A group of proteins that cause cell wall loosening during cell expansion.

EXTENSIN A protein in the primary cell wall that locks the cellulose fibers in place after cell growth.

FIBER An elongated cell with pointed ends, a thick secondary wall, and a small lumen. It is nonliving and commonly occurs as a group, and its function is of mechanical support. It is one of the two types of cells that make up sclerenchyma.

FIBER TRACHEID A type of cell that is intermediate in size and structure to a tracheid and libriform fiber. It has a secondary cell wall and is dead at maturity.

FIBROUS ROOT SYSTEM Consists of a shallow root system with limited growth of the primary root but with several lateral roots of similar length.

FILIFORM APPARATUS Finger-like ingrowths of cell wall in a cell. Commonly found in synergids, and it facilitates the entry of a pollen tube into an embryo sac.

FUNDAMENTAL TISSUE See *Ground tissue*.

FUSIFORM INITIALS One of the two types of cells that form the vascular cambium. They are elongated cells with pointed ends and produce secondary xylem toward the inside and secondary phloem toward the outside in a stem or root.

GAMETOPHYTE The gamete (egg and sperm)-producing structure or a phase in the life cycle of a plant; is haploid (1n).

GLUME Leaf-like structure that encloses a spikelet.

GROUND MERISTEM A group of cells in the apical meristem that produce the ground (fundamental) tissue.

GROUND (OR FUNDAMENTAL) TISSUE A tissue in an organ that is other than dermal and vascular tissues. It has several functions, including metabolism, photosynthesis, storage, and in some cases support of a structure.

GROWTH RING Ring-like structure in a wood formed by differences in seasonal activity of the vascular cambium resulting in xylem cells with differences in cell diameter and cell wall thickness. It has been called an annual ring and is used for determining the age of a tree, but there can be more than one ring in a year because of environmental changes during a season.

GUARD CELLS A pair of bean-shaped cells that regulate the opening and closing of stomata; commonly found in a leaf or stem.

HALOPHYTE Plant that lives under conditions of high salinity.

HAPLOID EMBRYO Embryo produced from a haploid cell(s) (e.g., microspore or immature pollen grain).

HAUSTORIA (SINGULAR = HAUSTORIUM) Absorbing structures produced by parasitic plants that invade the host and draw nutrition from it.

HEARTWOOD Interior part of the wood that is no longer active in the transport of water but provides mechanical strength.

HESPERIDIUM A type of fleshy fruit that develops from superior ovary and has a leathery exocarp (e.g., orange).

HETEROBLASTY A condition when in a plant there are leaves of different form at different developmental stages.

HETEROPHYLLY More than one type (form) of leaf in a plant.

HETEROSPORY Two types of spores produced by a plant; microspore (small) and megaspore (large).

HISTOGEN A layer of meristematic cells in the root or shoot apical meristems.

HOMOSPORY Spores of the same size and kind produced by a plant.

HYDROPHYTE Plant that lives in water or under very wet conditions.

HYPANTHIUM A floral tube formed by the fusion of sepals, petals, and stamens in a flower; the carpels are located within the tube.

HYPOCOTYL Stem-like structure below the cotyledons in dicotyledons; it extends until the primary root.

HYPODERMIS A layer of cells immediately below the epidermis; usually composed of sclerenchyma or collenchyma.

HYPOGENOUS A flower that has all floral organs attached below the ovary (i.e., ovary is located above them, also called superior ovary).

HYPOPHYSIS The uppermost cell of the suspensor that in some cases participates in embryo development.

IDIOBLAST An individual cell that is different from the cells surrounding it.

INFLORESCENCE A collection of flowers arranged in a cluster or as a group.

INTEGUMENT Outer one or two layer(s) of tissue of an ovule.

INTERCALARY MERISTEM The meristem that is located between maturing tissues above and below them.

INTERNODE The region between two successive nodes on a stem.

INTINE The inner cell wall of a pollen grain. It is thinner than the exine (outer wall) and has a role in the formation of pollen tube.

LAMINA Another name for leaf blade.

LATERAL ROOT Branch of a root that is produced from another root. It has its origin in the pericycle.

LATICIFER A tube-like secretory structure in a stem. It can be unicellular or multicellular and produces a milky substance called latex.

LEAF GAP An area above a leaf trace in a node without the vascular tissue that is usually occupied by parenchyma.

LEMMA Modified leaf-like structure in a grass flower.

LENTICEL An area of loose parenchyma cells produced by the cork cambium to the outside, instead of cork. It provides aeration to inner tissues of the stem.

LEUCOPLAST A colorless plastid that may contain starch, oil, or protein.

LIBRIFORM FIBER A cell that has a thick secondary wall and a small lumen, has pointed ends, and is dead at maturity. Its function is to provide mechanical support.

LIGNIN A complex polymer of aromatic alcohols that fills the space between cellulose fibers in a secondary cell wall, making it a tough wall.

LODICULE A swollen structure at the base of grass flower. It helps in the opening of the flower.

MERISTEM A group of cells that have the inherent ability for continued cell division.

MESOCOTYL The first internode located between the scutellum and coleoptile in a monocot embryo.

MESOPHYLL The ground tissue in a leaf blade located between the upper and lower epidermis.

METAPHLOEM The later-maturing phloem elements.

METAXYLEM The later-maturing xylem elements.

MICROSPORE Haploid structure produced after meiosis during microsporogenesis; develops into a male gametophyte (pollen grain in gymnosperms and angiosperms).

MIDDLE LAMELLA A layer made of mostly pectins that holds the cell walls of adjoining cells.

MONOECIOUS Plant that produces separate male and female flowers at different locations on a stem (e.g., in corn).

MULTILACUNAR NODE A node with several leaf gaps formed above the vascular traces to a leaf.

MYCORRHIZAE Root with associated fungi. If the fungus forms a mat on the surface it is ectomycorrhiza, and if it penetrates the root it is endomycorrhiza.

NECTARY A glandular structure found in flowers and some vegetative organs that produces nectar, a sugar-rich secretion.

NUCELLUS A diploid tissue located inside an ovule; produces megaspores after meiosis.

OVULE A female reproductive structure located inside an ovary; houses the embryo sac that contains an egg.

PALEA A leaf-like structure in a grass flower produced in place of a petal.

PALISADE PARENCHYMA Parenchyma cells located immediately below the upper epidermis (adaxial surface) in a leaf blade. These cells are upright and are the site of most of the photosynthesis in a leaf.

PALMATE VENATION Arrangement of veins in a leaf blade with three or more veins running from the base to the margins of leaf, like fingers in the palm of a hand.

PANICLE A branched raceme inflorescence (see *Raceme*).

PARALLEL (OR STRIATE) VENATION Veins in a leaf blade that run parallel from the base to the tip of a leaf.

PARENCHYMA A cell type or a tissue consisting of cells of different shapes. They have a thin primary cell wall, are living, and perform a variety of functions, including metabolism, photosynthesis, storage, and secretion of substances.

PERENNIALS Plants that grow for many years.

PERFORATION PLATE Opening at the two longitudinal ends of a vessel element.

PERIANTH Collective term for sepals and petals of a flower.

PERICARP Fruit wall that develops from the ovary wall.

PERICLINAL CELL DIVISION Cell division that is parallel to the outer surface of a structure.

PERICYCLE One or more layer of cells that surround the vascular tissue in a root. It consists of a layer(s) of parenchyma or sclerenchyma and has many functions, including the formation of lateral roots, shoot buds, and vascular cambium in a root.

PERIDERM Secondary dermal tissue produced in a stem or root by the cork cambium (phellogen) in plants with secondary growth. It includes the cork cambium, cork (phellem), and phelloderm.

PERIGYNOUS A flower in which the hypanthium does not fuse with the ovary and other floral organs arise from the edge of a cup-like hypanthium.

PETALS Nonreproductive colored organs of a flower that are produced inner to sepals.

Glossary

PETIOLE The stalk that connects a leaf blade to the stem.

PHELLEM (CORK) Produced by the cork cambium toward the outside of a stem. It consists of cells that are dead at maturity and contain suberin (a fatty substance) in their cell walls, which prevents the loss of water from the stem.

PHELLODERM Produced by the cork cambium toward the inside of a stem; consists mostly of parenchyma cells, sometimes called green parenchyma.

PHELLOGEN (CORK CAMBIUM) A meristem that produces the secondary dermal tissue (periderm) in plants with secondary growth.

PHLOEM A tissue that is a component of vascular tissue and is responsible for the transport of nutrients (e.g., sugars) and hormones in the plant body. In angiosperms, it normally consists of sieve tube elements, companion cells, fibers, and parenchyma cells.

PHRAGMOPLAST Formed during cell division, a ring of microtubules, vesicles, and endoplasmic reticulum. It contributes to the formation of a new cell plate.

PHYLLOTAXY The arrangement of leaves along the stem; there are different patterns found in plants.

PHYTOMERE A portion of a stem consisting of a leaf attached at a node, an internode, and an axillary bud.

PINNATE VENATION Pattern of veins in a leaf blade that are arranged like a pinna (feather-like).

PIT A thin area in cell wall with an intact cell membrane.

PIT FIELD Thin area in the cell wall where plasmodesmata are usually concentrated.

PLASMALEMMA (CELL MEMBRANE OR PLASMA MEMBRANE) A unit membrane that encloses the protoplast of a cell. It is mainly composed of proteins and phospholipids and has some carbohydrates and cholesterol molecules.

PLASMODESMATA Openings in plant cell wall that provide intercellular communication between adjoining cells.

PLASTIDS A class of pigment-carrying organelles (i.e., chloroplast and chromoplast) and storage-carrying organelle (amyloplast).

POLAR NUCLEI Nuclei in the central cell of an angiosperm embryo sac, commonly two in number but can be up to five. They fuse with a sperm and form the primary endosperm nucleus (commonly 3n).

POLLEN GRAIN Male gametophyte in angiosperms and gymnosperms that contains, or produces, sperm cells.

POLYARCH A condition of primary xylem in a root with many strands, each with protoxylem at the tip and alternating bands of phloem. Commonly found in monocotyledons.

POME A type of fleshy fruit developed from an inferior ovary (e.g., apple).

PRIMARY BODY Plant body produced by the shoot and root apical meristems.

PRIMARY CELL WALL The cell wall that encloses the protoplast of all plant cells. It is made up of cellulose, hemicelluloses, pectins, protein (extensin), and water.

PRIMARY ENDOSPERM NUCLEUS Nucleus formed by the fusion of sperm and polar nuclei; gives rise to the endosperm. It is commonly 3n but can be 4n or 5n.

PROCAMBIUM Formed from the provascular tissue, they are elongated cells and produce vascular tissue in plant organs.

PROMERISTEM A group of cells in a meristem that is self-perpetuating with a potential for indefinite cell division.

PROTODERM A layer of cells produced from the shoot and root apical meristem that gives rise to the dermal tissue in plant organs.

PROTOPHLOEM The first-maturing phloem elements in a plant organ.

PROTOPLAST The functional unit of a cell excluding the cell wall.

PROTOSTELE A type of stele with a central core of xylem surrounded by phloem. Found in early vascular plants (e.g., *Psilotum*).

PROTOXYLEM The first-formed mature xylem elements.

PROVASCULAR TISSUE A ring of small densely stained cells in the apical meristem; the precursor to procambium.

PULVINUS A swollen structure at the base of a leaf petiole that has a role in leaf movement.

QUIESCENT CENTER A group of cells in the center of root apical meristem that divide relatively less frequently than the surrounding cells of the meristem.

RACEME A type of inflorescence with flowers produced on an axis; each flower has a stalk, older ones at the base and the young ones at the tip (e.g., *Arabidopsis* and *Brassica*).

RADICLE The first young root that emerges from a seed.

RAY INITIALS One of the two types of cells that make the vascular cambium. They are small cells and produce parenchyma toward both the inside and outside of a stem or root.

RETICULATE VENATION A type of venation that forms a net-like pattern in a leaf blade. It has a large main vein that branches into smaller veins, ultimately ending in small veinlets.

RHIZOME An underground stem that generally spreads and produces new shoots.

RHIZOSPHERE An area that surrounds the root and has many microorganisms (e.g., fungi and bacteria).

RHYTIDOME The outer part of bark, which includes outer cork and any other nonfunctional tissues.

RING POROUS WOOD The wood in which there are alternating bands of large (early wood) and small (late wood) vessels (pores).

ROOT APICAL MERISTEM (RAM) A group of cells at the tip of the root that are ultimately responsible for producing the entire root system.

ROOT CAP A group of cells produced by the root apical meristem toward the tip. They have a variety of functions, including protecting the meristem, penetrating the root into the ground, and a role in gravitropism.

ROOT HAIR An outgrowth of a root epidermal cell, it is unicellular and has a direct role in the absorption of water and minerals from soil.

ROOT NODULE Swelling in a root caused by the invasion of a bacterium (*Rhizobium* sp.).

SAMARA A type of fruit that is nondehiscent and has a single seed and wings (e.g., maple).

SAPWOOD The recently formed part of the wood that is active in water transport.

SCLEREID A type of cell with a thick secondary wall, it may be living or nonliving, may occur singly or in groups, and has a variety of shapes. It provides support to the structure. It is one of the two types of cells of sclerenchyma.

SCLERENCHYMA A tissue, but can occur as a single cell or in a small group. Cells are of two types, fibers and sclereids (see separate definitions), and are mostly nonliving but can be living. The cells have a thick secondary wall with a small lumen, and they provide mechanical support.

SCUTELLUM The single cotyledon in a monocot embryo.

Glossary

SECONDARY BODY Growth in a stem or root produced by vascular cambium and cork cambium.

SECONDARY CELL WALL Cell wall that is formed in some plant cells and is laid inner to the primary wall. It consists of a high concentration of cellulose fibers with lignin deposited between them. The wall provides mechanical support to the cell.

SELF-INCOMPATIBILITY When the pollen from a flower of a species cannot effect fertilization of an egg in the flower of the same species (i.e., lack of self-fertilization).

SEPAL A member of the outermost whorl of organs in a flower; it is leaf-like and green.

SESSILE A leaf or a floral organ lacking a stalk.

SHOOT APICAL MERISTEM (SAM) A group of cells at the tip of a shoot that are ultimately responsible for producing the entire shoot system.

SIEVE AREAS Openings on the sides of a sieve tube element or sieve cell for lateral movement of substances.

SIEVE CELL A cell type of phloem in gymnosperms. Cells are elongated with tapered ends and have primary cell wall and sieve areas on the sides, but no sieve plate.

SIEVE PLATE End wall of a sieve tube element, connecting the cell above and below; has pores surrounded by callose.

SIEVE TUBE ELEMENT A cell type in the phloem of angiosperms; has a primary cell wall and living cytoplasm but no nucleus. It has a role in the transport of organic solutes (i.e., sugars and hormones).

SILIQUE A type of fruit that is dehiscent and develops from two carpels with two locules that are separated by a central partition (e.g., *Brassica* sp.).

SIPHONOSTELE A type of stele with a central pith surrounded by vascular tissue. Found in the stem of some ferns.

SOMATIC EMBRYO Embryo produced from any somatic (body) cell(s).

SPIKE A type of inflorescence with flowers produced singly and without a stalk (i.e., directly on an axis), with older ones at the base and young ones at the tip (e.g., wheat).

SPONGY PARENCHYMA Parenchyma cells below the palisade parenchyma in a leaf blade. Cells are commonly of irregular shape and have large intercellular spaces.

SPORE Haploid structure produced by a sporophyte after meiosis; it develops into a gametophyte.

SPOROPHYTE The spore-producing structure or phase in the life cycle of a plant; it is diploid (2n).

SPOROPOLLENIN A complex polymer present in the exine of pollen grains. It makes the exine tough for pollen preservation and resistant to decay.

STAMEN Male reproductive organ in angiosperm flowers. It comprises an anther at the top and a stalk below, the filament.

STELE A column of vascular tissue in the center of a stem or root; may have some ground tissue associated with it.

STIGMA It is the tip of the style of a carpel where pollen grains land and germinate.

STOLON An aboveground horizontal stem that spreads and produces new shoots (e.g., strawberry).

STOMATA (SINGULAR = STOMA) Openings in the epidermis that allow for gas exchange and transpiration.

STORIED CAMBIUM A condition where the fusiform initials are relatively short in length and are arranged in horizontal rows.

SUBERIN A fatty substance present in the cell wall of cork cells and as a strip (Casparian strip) in the endodermis.

SUBSIDIARY CELLS Cells adjacent to guard cells of a stomata; believed to have a role in the closing and opening of stomata.

SUSPENSOR A filamentous structure that links the developing embryo to the parental tissue.

SYMPLAST The protoplasmic continuum between cells facilitated through plasmodesmata in the cell wall.

SYNCARPOUS A condition when all the carpels in a flower are fused.

SYNERGID A cell in an embryo sac, commonly two in number, through which pollen tube enters for fertilization. It is located next to the egg cell and is part of the egg apparatus.

TAP ROOT SYSTEM Consists of a deep penetrating primary root with several lateral roots.

TAPETUM A layer of tissue in an anther that surrounds the developing microspores and pollen grains. It provides nutrients and many essential metabolites for pollen development.

TENDRIL A modified leaf or stem that curls around a support structure.

TEPALS Structures in place of sepals and petals in a flower that are indistinguishable (e.g., lily).

TETRARCH Primary xylem of the root with four arches, each with protoxylem at the tip.

TETRASPORIC EMBRYO SAC A condition in which all four potential megaspores participate in the formation of an embryo sac.

THORN A modified shoot with a hard pointed tip.

TRACHEID A cell type in xylem that has a secondary cell wall, is nonliving, and has an empty space in the center. The cell is elongated with overlapping ends. Primary role is in the transport of water and minerals, and in mechanical support.

TRANSFER CELL Cell with inner projections of the cell wall and membrane that enhance transport of materials across the cell membrane.

TRANSMITTING TISSUE A tissue in the style of a carpel through which pollen tube travels to reach the ovary.

TRIARCH Primary xylem of the root with three arches, each with protoxylem at the tip.

TRICHOME (HAIR) An outgrowth from epidermal cell of a stem, leaf, or floral organ. It may be unicellular or multicellular, occurs in various shapes, and may have a role in secretion of materials.

TUBER Swollen underground stem used for propagation, as in potato.

TUNICA The outer layer(s) in the shoot apical meristem of angiosperms in which cell divisions are predominantly anticlinal. Commonly one or two layers but can be up to five layers.

UMBEL A type of inflorescence with flowers grouped at the tip of an axis, and each flower has a stalk of similar length (e.g., onion).

UNIFACIAL LEAF Leaf that lacks a distinct upper and lower surface (e.g., a tube-like leaf in onion).

UNILACUNAR NODE A node with a single leaf gap formed by a vascular trace to a leaf.

VASCULAR CAMBIUM Meristem that produces secondary vascular tissue in plants with secondary growth. It consists of two types of cells, fusiform initials and ray initials.

VASCULAR TISSUE Consists of xylem and phloem and has roles in transport of water, minerals, and organic solutes, and mechanical support in the case of xylem.

VEINLET The very fine small branch of a vein.

VELAMEN A multilayered epidermis in the aerial root of some orchids.

VESSEL ELEMENT A cell type in xylem with secondary wall; it is nonliving, with an empty space in the center and an opening (perforation plate) at the two longitudinal ends of the cell. Primary role is in the transport of water and minerals, and mechanical support. Vessel elements are connected end to end to form a long vessel that may run the entire length of a stem or a root.

VIVIPARY Precocious germination of a seed (i.e., without dormancy) while still attached to the parent plant.

XEROPHYTE Plant that lives under dry arid conditions.

XYLEM A tissue that is a component of vascular tissue and is responsible for transport of water and minerals in the plant body. In angiosperms, it commonly consists of vessel elements, tracheids, fibers, and parenchyma cells.

BIBLIOGRAPHY

André, Jean-Pierre. *Vascular Organization of Angiosperms: A New Vision*. Science Publishers, Enfield, NH, 2005.
Beck, Charles B. *An Introduction to Plant Structure and Development*. Cambridge University Press, Cambridge, UK, 2005.
Bedinger, P. The remarkable biology of pollen. *Plant Cell* 4: 879–887, 1992.
Bell, Adrian D. *Plant Form*. Oxford University Press, Oxford, UK, 1993.
Bracegirdle, Brian, and Miles, Patricia H. *An Atlas of Plant Structure*, Vols. 1 and 2. Heinman Educational Books, London, 1971.
Cutler, David F., Botha, Ted, and Stevenson, Dennis. *Plant Anatomy: An Applied Approach*. Blackwell Publishing, Malden, MA, 2007.
Cutter, Elizabeth G. *Plant Anatomy*. Part 1: Cells and Tissues. Edward Arnold Publishers, London, 1969.
Cutter, Elizabeth G. *Plant Anatomy*. Part 2: Organs. Edward Arnold Publishers, London, 1971.
Dickinson, William C. *Integrative Plant Anatomy*. Harcourt Academic Press, San Diego, 2000.
Esau, Katherine. *Plant Anatomy*. John Wiley & Sons Inc., New York, 1953.
Esau, Katherine. *Anatomy of Seed Plants*, 2nd edition. John Wiley & Sons Inc., New York, 1977.
Evert, Ray F. *Esau's Plant Anatomy*, 3rd edition. John Wiley & Sons Inc., Hoboken, NJ, 2006.
Fahn, A. *Plant Anatomy*, 4th edition. Pergamon Press, Oxford, UK, 1990.
Fosket, Donald E. *Plant Growth and Development: A Molecular Approach*. Academic Press Inc. San Diego, CA. 1994.
Greyson, Richard I. *The Development of Flowers*. Oxford University Press, Oxford, UK, 1994.
Howell, Stephen H. *Molecular Genetics of Plant Development*. Cambridge University Press, Cambridge, UK. 1998.
Maheshwari, P. *An Introduction to the Embryology of Angiosperms*. McGraw-Hill, New York, 1950.

Mauseth, James D. *Plant Anatomy*. Benjamin/Cummings Publishing Co. Inc., Menlo Park, CA, 1988.

O'Brien, T. P., and McCully, Margaret E. *Plant Structure and Development: A Pictorial and Physiological Approach*. McMillan Co., Toronto, 1969.

Peterson, R. Larry, Peterson, Carol A., and Melville, Lewis H. *Teaching Plant Anatomy: Through Creative Laboratory Exercises*. National Research Council of Canada, Ottawa, 2008.

Rudall, Paula. *Anatomy of Flowering Plants: An Introduction to Structure and Development*, 2nd edition. Cambridge University Press, Cambridge, UK, 1992.

Shivanna, K. R. *Pollen Biology and Biotechnology*. Science Publishers Inc., Enfield, NH, 2003.

Steeves, Taylor A., and Sussex, Ian M. *Patterns in Plant Development*, 2nd edition. Cambridge University Press. Cambridge, UK, 1989.

Tucker, S. Floral development in legumes. *Plant Physiology* 131: 911–926, 2003.

Yadegari, R., and Drews, G. N. Female gametophyte development. *The Plant Cell* 16 (Suppl.): S133–S141, 2004.

INDEX

Page references for figures are indicated by *f* and for tables by *t*.

abaxial surface, 20, 95
abscisic acid (ABA): leaf abscission, 110; leaf form, 105; seed dormancy, 53
abscission, leaf, zone, 110
achene, 54*t*
acropetal, 68
adaxial surface, 20, 95
Adiantum: root apical meristem, 116*f*
adnation, 22
adventitious roots, 113, 124–125
adventitious shoot buds, 32, 66
aerenchyma, 71, 72*f*, 120
aerial roots, 113, 119
aggregate-accessory fruit, 54*t*, 55
aggregate fruit, 54*t*
aleurone grains, 12
Allium cepa: female gametophyte, 37–38; root apical meristem, 117*f*
alternate phyllotaxy (leaf arrangement), 58, 58*f*
alternation of generations, 4–5, 5*f*, 31, 34, 42
Amelanchier alnifolia (Saskatoon berry): flower, 25–26, 26*f*
amyloplasts, 10, 11–12

anatomy. *See also specific topics*: comparative, 1, 5; definition, 1; developmental, 5–6; utilitarian purposes, 1
androecium, 20
Anemone patens (prairie crocus): flower, 24, 25*f*, 36*f*
angiosperms, 2; alternation of generations, 5, 42; vegetative reproduction, 32–34; sexual reproduction, 34–43
annual plants, 17
annual rings. *See* growth rings
annular tracheary elements, 74, 75*f*
anther, 20, 20*f*
antheridia, 5*f*
anticlinal, cell division, 47, 87, 107, 123–124, 131–132; shoot apical meristem, 61, 62
antipodal cells, 21, 37, 37*f*. *See also* embryo sac
apetalous, 22
apical dominance, 66
apocarpous ovary, 21
apomixis, 42
apopetaly, 22
apoplast, 14

aposepaly, 22
Arabidopsis thaliana (Arabidopsis), 19, 21, 23; embryo development, 45; pollen grain, 40f; pollen development, 39f; root apical meristem, 116
archegonium, 5, 5f, 34
asexual reproduction. *See* vegetative reproduction
astrosclereid, 73, 73f
atactostele, 93f, 94
auxins, 66; apical dominance, 66; leaf abscission, 110; root formation, 124–125
Avena sativa (oats): inflorescence, 19
awn, 26, 27f
axillary bud, 57, 66
axillary meristems, 61, 61f, 65–66

banyan tree (*Ficus benghalensis*): aerial roots (prop roots), 113
bark, 140, 140f
basal cell, 45f, 46
basipetal, 85–86, 89–91
berry, fruit type, 54t
bicollateral vascular bundle, 83–84, 84f
biennial plants, 17
bilateral symmetry, 47
bisporic embryo sac (female gametophyte), 38
bordered pits, 74, 75f, 80
brachysclereid, 73f
bract, 18f, 22
branching: lateral, 65; shoot, 65–66; terminal, 65
Brassica napus (canola): flower morphology, 24, 25f; inflorescence, 19
buds: adventitious shoot, 32, 66; axillary, 57, 66; shoot buds, 125–126, 125f
bud scales, 64, 64f, 100
bulbs, 33; leaf, storage organs, 100
bundle cap, 82f, 83
bundle sheath, 103–104, 104f; C_4 photosynthesis, 104

callase, 38
callose (β-1, 3 glucan), 38, 39f, 77
calyptrogen, 116
calyx, 20, 20f

cambium: stratified (storied), 131–132, 131f; vascular, 3f, 4, 128–133. *See also* vascular cambium
Cannabis, dioecious plant, 22
canola (*Brassica napus*): floral organ initiation, 29; flower, 24, 25f
capitulum (Head), inflorescence, sunflower, 18f, 19
Capsella bursa-pastoris: embryo development, 45–49, 45f–49f; endosperm formation, 52
capsule, fruit type, 54t
carpels, 20f, 21; development, 29; growth, 30; pollination, 40
caryopsis, 54t
Casparian strip, 103, 104f, 120, 121f; suberin, 120, 141
cataphylls, 64, 64f, 100
cell culture, 51–52
cell division, shoot growth, 64–65
cell enlargement, elongation, 16, 64–65
cell growth, 16
cell membrane, 8–9, 8f
cell structure, 7–16. *See also specific components*: cell wall, 8f, 13–15, 14f, 15f; definition, 7; diagram, 8f; discovery, 7; growth, 16; protoplast, 7, 8–13; shape, 7–8
cell suspension cultures, 34, 51–52
cell types, 69–80. *See also specific tissues and structures; specific types*: aerenchyma, 120; chlorenchyma, 70; collenchyma, 71–72, 71f, 83; cork, 80, 139f; epidermal, 71f, 78–80, 79f; parenchyma, 70–71, 71f, 72f (*see also* palisade and spongy parenchyma (mesophyll)); sclerenchyma, 53, 72–73, 73f, 83; secretory structures, 80; sieve cell, 77–78; sieve elements, sieve tube elements, 76–78, 77f; tissue systems, 69–70; tracheary elements, 73–76, 74f, 75f
cellulose: cell growth, 16; cell wall, 13; secondary cell wall, 15, 15f
cell wall, 7, 8f, 13–15, 14f, 15f; cellulose, 13; cytokinesis, 13; exoskeleton, 13; extensin, 13; hemicellulose, 13; microfibrils, 13; middle lamella, 8f, 13; pectin, 13; phragmoplast, 13, 14f;

pits, 15, 15f; plasmodesmata, 14; primary pit field, 8f, 14; primary wall, 8f, 14; secondary wall, 14–15, 15f; wall matrix, 13
central cells, embryo sac, 21, 36, 37f; endosperm formation, 52; fertilization, 41, 42f
central zone: 61–62, 62f; reproductive shoots, 68
chalazal end, embryo sac, 36, 36f
chlorenchyma, 70
chloroplasts, 8f, 10; chlorenchyma, 70; without grana, 104; starch grains/granules, 11–12, 12f
chromoplasts, 10
Chrysanthemum: inflorescence, 28
Citrus, apomixis, 43
cladophylls, 66–67
coconut: endosperm, 56
coleoptile, 50, 50f, 60
coleorhiza, monocotyledon embryo, 50, 50f
Coleus: shoot apex, 61f; stem, 84f
Collateral vascular bundle, 82f, 83
collenchyma, 71–72, 71f; stem, 83
companion cells, 77f, 78
complementary cells, 140, 140f
compound leaves, 96, 96f, 97, 100f; development, 108
conifer wood, 133, 132f
cork cambium, 3f, 4, 121, 139f; periderm, 139–140, 139f
cork cells, 80, 139, 139f. See also phellem
corm, 33
corolla, 20, 20f
corpus, shoot apical meristem, 61, 61f
cortex, stem, 82, 82f, 83; root, 119–120
corymb, inflorescence, 18f, 19
cotyledon: development, dicotyledons, 47–48, 48f, 59–60; monocotyledon, 49. See also scutellum
cross-pollination, 40–41
crystals, 12, 12f, 137
Cucurbita: stem, 84f
Cucurbitaceae: flower morphology, 22
cuticle, 78; leaf, 101; root, 119; stem, 83
cutin, 78
cytokinesis, 13, 38
cytokinins, 66, 111

cytoplasm, 8f, 9
cytoskeleton, 11

dandelion *(Taraxacum)*, apomixis, 43
dehiscent fruit, 54t, 55
dermal tissue (epidermis), 69; cork cells, 80, 139f; dicotyledons, 82–83, 82f; root, 118, 118f, 119
detached meristems, 61f, 65–66
diarch xylem, 121, 122f
dicotyledons. See also specific topics: embryo development, 45–49, 45f–49f; leaf form, 95, 95f; stem tissue, 82–84, 82f, 84f. See also root tissues; stem tissues
dictyosome, 8f, 9
dictyostele, 93f, 94
diffuse porous wood, 134, 135f
dioecious, 22
diploid, 4, 5f; sporophyte, 31; zygote, 5
dormancy: embryo, 53–54; seed, 49, 54–55
double fertilization, 41, 42f
double haploids, 51–52
drupe, fruit type, 54t, 56
druses, 12, 12f
dry fruit, 54t, 55
dwarf mistletoe: haustoria, 114

ectomycorrhizae, 113f, 114
egg, 4, 5f, 21, 36, 37f; fertilization, 41, 42f; seed plants, 35
egg apparatus, 36, 37f. See also embryo sac
elaioplasts, 10, 12
embryo, 4; definition, 44; development, 45–51; haploid, 51; meristems, 3; somatic, 51–52
embryo development, 44–51; angiosperms, variation, 44–45; dicotyledons, 45–49, 45f–49f; monocotyledons, 49–51, 50f; suspensor, 51
embryogenesis: patterns, 44–51. See also embryo development
embryo sac (female gametophyte), 35, 36f, 37; fertilization, 41, 42f; *Fritillaria* type, 38; *Plumbago* type, 38; *Polygonum* type, 38. See also antipodal cells; central cell; egg apparatus; polar nuclei; synergids
endarch, 122
endocarp, 56

endodermis, 120, 121f, 124f
endomembrane system, 8f, 9
endomycorrhizae, 114
endoplasmic reticulum (ER), 8f, 9
endosperm, 41–42, 52–53; cellular, 52; formation, 52–53; *Fritillaria*, 42; monocotyledons, 50, 50f; *Polygonum* type, 42; primary endosperm nucleus, 41–42, 42f, 52
endothecium, 20
epicotyl, 59–60
epidermal cells, 71f, 78–80, 79f. *See also* epidermis
epidermal hairs (trichomes), 79, 79f; secretory, 80
epidermis, 71f, 78; anther, 20; stem, 82, 82f, 83, 88, 91, 92f; guard cells, 78–79, 79f; hairs, 79, 79f; leaf, 101–102, 102f; root, 118, 118f, 119; secretory, 80; stomata, 78, 79f
epigeal seed germination, 59
epigynous, 22, 23f, 26, 26f
ergastic substances, 11–12, 12f
ethylene, leaf abscission, 110
Euphorbia sp.: asexual reproduction, 32; roots, 111
eustele, 93f, 94
evolution: gamete fusion and meiosis, 31
exarch, 121–122
exine, 21, 38, 39, 40f
exocarp, 56
exodermis, 120
exoskeleton, 13
expansins, 16
extensin, 13, 16

female gametophyte, 21; development and structure in angiosperms, 35–38, 36f, 37f; *Selaginella*, 34–35
ferns: alternation of generations, 5; life cycle, 4–5, 5f; sexual reproduction, 34
fertilization, 40–42, 42f; double, 41, 42f; fern, 4
fibers, 72–73; bundle caps, 83; cellulose, 15; lamina, 103; libriform, 135; phloem, 72, 78, 82, 84; secondary phloem, 137, 137f; secondary xylem, 135, 136f; stem, 83, 84, 89, 91; xylem, 72, 74, 78

fiber tracheid, 135
fibrous root system, 112f, 113
filament, stamen, 20, 20f; growth, 30
filiform apparatus, 36–37, 42f; fertilization, 41, 42f
fleshy fruit, 54t, 55
floral apex, 27–29, 28f
florets, 18f, 19; Poaceae, 26
flower, 17–19; development, 29–30; flowering, induction, 27–29, 28f; formation, function, position, 17–18; inflorescence, 18–19, 18f; uses, 17; variation, 19, 20f, 22–27, 23f–27f
flowering shoots, 68
flower morphology, 19–27, 20f; anther, 20, 20f; apetalous, 22; bisexual, 22; bract, 18f, 22; Brassicaceae, 24, 25f; calyx, 20, 20f; carpel, 20f, 21; corolla, 20, 20f; Cucurbitaceae, 22; dioecious, 22; epigynous, 22, 23f, 26, 26f; gynoecium, 21; hypogenous, 20f, 22, 24, 25f; incomplete, 22; involucre, 22; lodicule, 22, 26–27, 27f; monoecious, 22; nectaries, 22; ovary, 20f, 21; perianth, 20; perigynous, 22, 24f, 26, 26f; petals, 20, 20f; pistil, 21; Poaceae, 22, 26–27, 27f; pollen grains, 21, 40f; Ranunculaceae, 24, 25f; receptacle, 19, 20f; Rosaceae, 25–26, 26f; sepals, 19–20, 20f; stamens, 20, 20f; stigma, 20f, 21; style, 20f, 21; unisexual, 22; variations, 19, 22–27; whorls, 22
follicle, fruit type, 54t
Fraxinus Americana (ash): vascular cambium, 130f; wood, 136f
Fritillaria: endosperm, 42; female gametophyte, 38
fruit: classification, 54t, 55; covering, seed dispersal, 56; definition, 55; development, 55–56; pericarp, 55–56
fruit types, 54t; aggregate, 54t; aggregate-accessory, 54t, 55; dehiscent, 54t, 55; dry, 54t, 55; fleshy, 54t, 55; multiple, 54t; non-dehiscent, 54t, 55; simple, 54t, 55
fundamental system (tissue). *See* ground tissue
fusiform initials, 129–131, 130f, 131f; secondary xylem (wood), 133
fusion nucleus, 37f

Index

gametangia, 5
gamete, 4
gametophyte, 4, 5f, 21, 40f; haploid, 31; mosses, 34. *See also* male gametophyte, female gametophyte
gibberellins, root, 111
glossary, 143–153
glucoronoarabinoxylan, 14
glumes, 26
glyoxysomes, 11
Golgi body, 8f, 9
grasses: cell wall, 14; leaf development, 110; leaf form, 96, 97, 97f; floret, 26–27, 27f
ground meristem, 63, 63f; embryo, 47, 48f; root, 118; stem, 63f, 86–87, 87f, 88
ground tissue (fundamental tissue), 69, 70; dicotyledon stem, 82, 82f; formation, 47, 48f; monocotyledon stem, 91, 92f; root, 118, 118f, 119–120
growth rings (annual rings), 132–133, 132f; secondary xylem, 134
guard cells, 78–79, 79f, 101, 102, 102f
gymnosperms, alternation of generations, 5; shoot apical meristem, 62
gynoecium, 21

hairs (trichomes): epidermal, 79, 79f; root, 119; secretory, 80; stem, 83
halophytes: definition, 105; leaf form, 106, 106f
haploid, 4, 5f; double, 51–52; embryos, 51; fern life cycle, 4, 5f; gametophyte, 31; megaspores, 35, 36f; microspores, 21, 38, 39f
haustoria, 52, 115
head (capitulum), 18f, 19
heartwood, 133
Helianthus annuus (sunflower): inflorescence, 19; shoot apex, 62f; stem, 82f
helical phyllotaxy, 59f, 60, 60f
helical tracheary elements, 74, 75f
hemicelluloses: cell growth, 16; cell wall, 13
hespiridium, 54t
heteroblasty, 97
heterophylly, 97
heterospory, 34
hilum, 12
histogen, 115
homospory, 34

hydrophytes: definition, 105; leaf form and tissues, 105, 105f
hypanthium: cup-shaped, 23, 24f, 25, 26f; fused, 22, 23f; nonfused, 22, 24f
hypocotyl, 47; elongation, early, 59
hypodermis, 83
hypogeal seed germination, 59–60
hypogenous, 20f, 22, 24, 25f
hypophysis cell, 46

idioblasts, 73
incomplete flower, 22
indeterminate inflorescences, 18–19, 18f
inflorescense, 68; determinate, 18, 18f; indeterminate, 18–19, 18f; types, 18, 18f, 19. *See also* flower
inflorescence apex, 27–28, 28f; development, 29
insectivorous plants: digestive glands, 80, 100; leaf traps, 100
integuments, 21, 35, 36f; seed maturation, 53
intercalary meristem, 65
intercellular spaces, 71, 72, 72f
interfascicular parenchyma, 82, 128
intermediate filaments, 8f, 11
internodes, 57, 81; elongation, 64–65, 64f
intine, 38, 39, 40f
involucre, 22

Kalanchoe (Mexican hat): asexual reproduction, 33
Kranz (wreath) anatomy, 104

lamina (leaf blade), 95–97, 96f, 101–104, 102f, 103f
lateral roots, 112, 112f, 122f, 123–124, 124f
latex, 80
laticifers, 80
layering (asexual reproduction), 32, 33
leaf: bundle sheath, 103–104, 104f; C_3, C_4 photosynthesis, 104; photosynthesis, 95, 102; phyllotaxy, 58, 58f, 59f, 95; tissue organization, 95–96
leaf abscission, 110; role of ABA, auxin and ethylene, 110
leaf arrangement. *See* Phyllotaxy
leaf blade (lamina), 95–97, 96f, 101–104, 102f, 103f

leaf development, 106–110, 107f, 108f; compound leaves, 108; expansion, 109; grasses, 110; monocotyledons, 109; palms, 110

leaf form, 96–100, 96f; 104–105; blade (lamina), 95–97, 96f, 101–104, 102f, 103f; bundle sheath, 103–104, 104f; compound, 96, 96f; dicotyledon, 95, 95f; grass, 96, 97, 97f; halophytes, 106, 106f; heteroblasty, 97; heterophylly, 97; hydrophytes, 105, 105f; leaf sheath, 96, 97f; modifications, 104–106, 105f, 106f; monocotyledon, 95, 96f, 97–98; petiole, 96, 96f, 97; role of ABA, 105; sessile, 96; simple, 96, 96f, 97; stalk, 96, 96f; stipule, 96, 96f, 97; xerophytes, 104f, 105–106

leaf gap, 93

leaflets, 96, 96f

leaf primordium, 107, 107f; monocotyledon embryo, 50, 50f; shoot apical meristem, 61, 61f, 95

leaf tissues: blade (lamina), 95–97, 96f, 101–106, 102f, 104f–106f; halophytes, 106, 106f; hydrophytes, 105, 105f; petiole, 101; xerophytes, 104f, 105–106

leaf traces: dicotyledons, 84–86, 85f, 87f; monocotyledons, 91, 92f

leaf types, 99–100; alternate arrangement, 58, 58f; bud scales (cataphylls), 64, 64f, 100; compound, 96, 96f, 97, 100f; insect traps, 100; opposite arrangement, 58, 58f; spines, 100; spiral arrangement, 59f. 60, 60f; tendrils, 100, 100f; whorled arrangement, 58, 58f

leaf venation: palmate, 96f, 97, 98; patterns, 96f, 98–99, 99f, 103; pinnate, 96f, 97, 98; reticulate, 98, 99f; striate (parallel), 96f, 98–99, 99f; vein connections, 99, 99f

legume, fruit type, 54t

lemma, 26–27, 27f

lenticels, 140, 140f

libriform fibers, 135

lignin, 15, 72, 73f

Lilium female gametophyte, 38

Linum usitatissimum (flax): shoot apex, 63f

lodicule, 22, 26–27, 27f

long shoot, 57–58, 63–64, 64f; nodes and internodes, 81–82

Lupinus albus (lupin): shoot apex, 60f

Lycopodium: shoot branching, 65

maize. See *Zea mays*

male gametophyte: development and structure, 38–40, 39f, 40f; seed plants, 35; *Selaginella*, 34, 35. See also pollen

marginal meristem: leaf, 107, 108f

matrix, cell wall, 13

Medicago sativa (alfalfa): vascular cambium, 128f

megasporangium, 35

megaspore, 21; haploid, 35, 36f; mosses, 34–35; seed plants, 35

megaspore mother cell, 35, 36f

megasporocyte, 21

meiosis, 4, 5f, 21, 31, 35, 36f

meristem. See also *specific types*: detached (axillary), 61f, 65–66; embryonic, 3; intercalary, 65. See also root apical meristem; shoot apical meristem

mesocarp, 56

mesocotyl: monocotyledon embryo, 50, 50f

mesophyll, 102; palisade, 72, 102–103, 102f, 105f, 106f (*see also* palisade parenchyma); spongy, 102–103, 102f. See also spongy parenchyma

metaphloem, 84f, 90; root, 122, 122f

metaxylem, 84f, 90

microfibrils, 16; cell wall, 13

microfilaments, 8f, 11

micropyle, 35, 36, 37f; fertilization, 41, 42f

microspore: haploid, 21, 38, 39f; mosses, 34; seed plants, 35; tetrad, 38, 39f

microspore mother cells (MMCs), 20, 38

microsporocytes, 20

microtubules, 8f, 11

middle lamella, 8f, 13

mitotic spindle, 13

monocotyledons. See also *specific topics*: embryo development, 49–51, 50f; leaf development, 109; leaf form, 95, 96f, 97–98; secondary growth, 138; stem tissue, 91–92, 92f. See also stem tissues

monoecious, 22

mosses: sexual reproduction, 34–35

multilacunar node, 86

multiple fruit, 54t

Index

multiseriate rays, 135–136, 136f, 138
mycorrhizae, 113–114, 113f

nectaries, 22, 80
net-like (reticulate), venation, 98, 99f
nexine, 39
nitrogen fixation, root nodules, 114, 114f
nodes, 57, 81–82
nodules, root, 114, 114f
non-dehiscent fruit, 54t, 55
nucellus, 21, 35, 37f
nuclear envelope, 9–10
nuclear pore complexes, 9–10
nucleoli, 8f, 9
nucleoplasm, 9
nucleus, 8f, 9–10
nut, fruit type, 54t
Nymphaea (water lily); leaf tissues, 72f

opposite phyllotaxy (leaf arrangement), 58, 58f
Opuntia; cladophylls, 66
orthotropic leafy shoot, 57
osteosclereid, 73f
ovary, 20f, 21, 35; development, 29; inferior, 23, 23f; superior, 20f, 22
ovule, 21; development and structure, 35–38, 37f; fertilization, 41, 42f; pollination, 41; seed plants, 35

palea, 26–27, 27f
palisade parenchyma (mesophyll), 72f, 102–103, 102f, 105f, 106f; halophyte, 106, 106f; hydrophyte, 105f; leaf expansion, 108f, 109; leaf veins, 103; xerophyte, 106
palmate leaves, 96f, 97, 98
palmately compound leaf, 96f, 97
palmate venation, 96f, 97
palm tree: leaf development, 110
panicle, inflorescence, oats, 18f, 19
parallel (striate) venation, 96f, 98–99, 99f
parenchyma, 70–71, 71f, 72f; palisade (*see* palisade parenchyma); root, 119–120; secondary xylem, 135; secretory, 80; spongy (*see* spongy parenchyma); stem, 71f, 82, 83, 91
pear, sclereid, 15, 15f
pectic substances, 72
pectin, 13

pepo, fruit type, 54t
perennial plants, 17–18
perforation plates, 74f, 76. *See also* vessel elements
perianth, 20, 20f
pericarp, 55–56
periclinal, cell division, 47, 61
pericycle, 121, 121f, 123; lateral roots, 123, 124f; shoot buds, 126
periderm, 139–141, 139f, 140f; cork (phellem) cells, 80, 139f; cork cambium (phellogen), 139, 139f; phelloderm, 139, 139f
perigynous, 22, 24f, 26, 26f
peroxisomes, 10–11
petals, 20, 20f; development, 29–30; growth, 29–30
petiole, 96, 96f, 97, 101
Petunia: flower opening, 30
phellem (cork), 139, 139f
phelloderm, 139, 139f
phellogen (cork cambium), 4, 139, 139f
phloem, 2, 70; companion cells, 77f, 78; dicotyledons, 82–84, 82f, 84f; fibers, 72, 82, 84; formation, 47; leaf, 103, 104f; procambium, 63, 63f; root, 119, 121, 121f, 122, 122f; secondary, 136–138, 137f; sieve cells, 77, 78; sieve elements, 76–78, 77f; stem tissue, 89; stem tissue, basipetal extension, 90–91; vascular cambium, 128–133, 128f, 130f. *See also* secondary phloem
phospholipid layer, plasmalemma, 8f, 9
photosynthesis: C_4, 104; leaf, 95; stroma, 10
phragmoplast, 13, 14f
phyllotaxy, 58, 59f, 95; alternate, 58, 58f; leaf development, 106–107, 107f; opposite, 58, 58f; spiral (helical), 59f, 60, 60f; whorled, 58, 58f
phytomere, 57
pinnately compound leaf, 96f, 97
pinnate venation, 96f, 97
Pinus: wood, 132f
pistil, 21. *See also* carpels
pistillate flower, 22
pith, stem, dicotyledons, 82, 82f, 83
pits: bordered, 74, 75f, 80; cell wall, 15, 15f
pitted tracheary elements, 74, 74f, 75f

plant propagation. *See also specific types*: cell suspension cultures, 34, 51–52; tissue culture, 34
plasma membrane (plasmalemma), 8–9, 8f
plasmodesmata, 14
plastids, 8f, 10; starch, 11; storage lipids, 12
plate meristem, 107, 108f
Plumbago: female gametophyte, 38
Poaceae: flower morphology, 22, 26–27, 27f
polar nuclei, 21, 36; fertilization, 41, 42f;
pollen (pollen grain or male gametophyte), 20–21, 35, 38–42, 40f; development and structure, 38–40, 39f, 40f; germination, 40–41; seed plants, 35; three-celled, 21, 39, 40f; two-celled, 21, 39f
pollen sac, 38, 39f
pollen tube, 35, 40–41; fertilization, 41, 42f
pollen wall, 39–40, 40f
pollination, 40–41
polyarch xylem, 121, 122f
polygalacturonic acid, 13
Polygonum: female gametophyte (embryo sac), 21, 35–38, 37f, 42
pome, fruit type, apple, 54t
pore. *See also specific types*: pollen wall, 39–40; size in cell wall, 13
potato: tuber, 33, 33f
P-proteins, 76, 77, 89
prairie crocus (*Anemone patens*): flower, 24, 25f
primary body, 4
primary cell wall, 8f, 14
primary endosperm nucleus (PEN), 41–42, 42f, 52
primary root, 112–113
prismatic crystals, 12, 12f
procambium, 63, 63f; embryo, 47, 48f; leaf, 107, 108f; root, 119; secondary body, 128; stem tissue, 87f, 88–89, 91, 93f
promeristem, 62, 63, 63f; stem, 86, 87f
prop roots, 113
protoderm: embryo, 47, 47f, 48f; root, 118; shoot apex, 62–63, 63f; stem, 86–88, 93t
protophloem, 84f, 90; root, 122, 122f
protoplast, 7, 8–13, 8f; chloroplasts, 8f, 10; cytoplasm, 8f, 9; cytoskeleton, 11; dictyosome, 8f, 9; endoplasmic reticulum, 8f, 9; ergastic substances, 11–12, 12f; glycosomes, 11; ground plasm, 9; intermediate filaments, 8f, 11; microfilaments, 8f, 11; microtubules, 8f, 11; middle lamella, 8f; nucleus, 8f, 9–10; organelles, 8f, 9; peroxisomes, 10–11; plasmalemma, 8–9, 8f; plastids, 8f, 10; ribosomes, 8f, 9; starch grains/granules, 11–12, 12f; tonoplast, 8f, 11; vacuole, 8f, 11
protostele, 92–93, 93f; root, 119
protoxylem, 84f, 89–90, 121
provascular tissue, 62, 63f; root, 118; stem, 87–88, 87f, 93t
Pteridium (bracken fern): asexual reproduction, 32
pubescent, 20
pulvinus, 101

quiescent center: root, 118

raceme, inflorescence, 18f, 19
rachilla, 26, 27f
radicle, 112; monocotyledon embryo, 50, 50f
Ranunculaceae, flower morphology, 24, 25f
Ranunculus: root tissues, 118f
Raphanus sativus: root apical meristem, 117f
raphides, 12, 12f
ray initials, 129–131, 130f, 131f; secondary xylem (wood), 133
receptacle, 19, 20f
reproduction, 30–43. *See also specific types and topics*: alternation of generations, 31, 42; apomixis, 42; regeneration, 32; sexual, 34–42; sexual, overview, 42–43; vegetative (asexual), 32–34, 33f; zygote, 31
reproductive shoots, 67–68
residual meristem, stem, 86–87, 87f
resin canals: conifer wood, 132f, 133
resin ducts, 80
reticulate (net-like) venation, 98, 99f
rhamnogalacturonans, 13
rhizoid, 5f

rhizome, 5f, 32–33, 33f
rhizosphere, 123
rhytidome (outer bark), 140, 140f
ring porous wood, 134, 134f
Robinia psuedoacacia (locus): storied cambium, 131f
root: aerial, 113, 119; cytokinins, 111; functions, 111; gibberellins, 111; phloem, 119; primary, 112–113; prop, 113; protostele, 119; xylem, 119. *See also* adventitious root; lateral root
root apex, 116. *See also* root apical meristem (RAM)
root apical meristem (RAM), 3, 115–118; dicotyledon embryo, 48, 49f; meristem layers, 116, 117f; monocotyledon embryo, 50, 50f; open meristem, 116–118, 117f; single apical cell, 115, 116f
root associations, 113–115; haustoria, 52, 115; mycorrhizae, 113–114, 113f; nodules, 114, 114f
root branching, 123–125; adventitious, 113, 124–125; lateral, 112, 112f, 122f, 123–124, 124f
root cap, 46, 50, 115–117, 116f, 117f, 123, 124, 124f; monocotyledon embryo, 50
root hairs, 119, 120f
root nodules, 114, 114f
root shoot buds, 125–126, 125f
root systems, 2, 3f, 112–113, 112f; adventitious, 113, 124–125; fibrous, 112f, 113; tap root, 112, 112f
root tissues, 118–123; dermal (epidermis), 118, 118f, 119; differentiation pattern, 93f; ground, 118, 118f, 119–120; root cap, 115–117, 116f, 117f, 123; root hairs, 119, 120f; vascular, 118, 118f, 121–122, 121f, 122f
Rosaceae: flower morphology, 25–26, 26f
runner, strawberry, 33, 33f

samara, fruit type, 54t
sapwood, 133
Saskatoon berry (*Amelanchier alnifolia*) flower, 25–26, 26f
scale bark, 140
scalariform tracheary elements, 74, 75f
sclereids, 72–73, 73f, 80; pear, 15, 15f

sclerenchyma, 53, 72–73, 73f; leaf, 103, 104f, 106; petiole, 101; root, 120; secondary cell wall, 72, 73f; stem, 83, 88
scutellum, 50, 50f, 60
secondary body, 127–141; functions, 127; monocotyledons, 138; periderm, 139–141, 139f, 140f; phellogen, 4, 139, 139f; secondary growth, dicotyledons, 128–141; secondary phloem, 136–138, 137f; secondary xylem (wood), 133–136, 134f–136f; vascular cambium, 3f, 4, 128–133
secondary cell wall, 14–15, 15f; sclerenchyma, 72, 73f
secondary growth: dicotyledon, 128–141; monocotyledons, 138
secondary phloem, 136–138, 137f
secondary xylem (wood), 133–136, 134f–136f
secretory hairs, 80
seed: coat, 53; dispersal, 56; dormancy, 49, 53–55; food source, 17; habit, 35; proteins, 12; storage lipids, 12
seed development, 48–49, 49f, 52–55; embryo and seed dormancy, 53–55; endosperm, 41–42, 52–53; maturation, 53, 54t
seed dormancy, 53–55; role of ABA, 53
seed germination, 59–60
seedling, early growth, 59–60, 60f
seed plants. *See* angiosperms; gymnosperms
Selaginella, branching, 65
self-incompatibility, 40
self-pollination, 40
sepals, 19–20, 20f; development, 29; growth, 29
sessile, 18f, 19; leaf, 96
sex cells, 4
sexine, 39
sexual reproduction, 4, 5f, 34–42; alternation of generations, 4, 5f, 34, 42; angiosperms, 34; ferns, 34; fertilization, 41–42, 42f; mosses, 34–35; overview, 42–43; ovule and female gametophyte, 35–38, 36f, 37f; pollen and male gametophyte, 38–40, 39f, 40f; pollination, 40–41; seed plants, 35
shade leaves, 104
sheath, leaf, 96, 97f

shoot: flowering, 68; leaf distribution, 58, 58f; long, 57–58, 63–64, 64f; nodes and internodes, 57, 81–82; phyllotaxy, 58, 59f, 95; phytomeres, 57; reproductive, 67–68; short, 57–58, 64, 64f; upright (orthotropic) leafy, 57
shoot apex, 27–28, 28f, 60–63, 60f–63f; *Coleus*, 61, 61f; ground meristem, 63, 63f; *Helianthus*, 62f; leaf development, 95, 106–107, 107f; *Linum*, 63f; *Lupinus*, 60f; procambium, 63, 63f; promeristem, 62, 63, 63f; protoderm, 62–63, 63f; provascular tissue, 62, 63f
shoot apical meristem (SAM), 3, 27, 60–63, 60f–63f; central zone, peripheral zone, radial zonation, 61–62, 62f; embryo dicotyledon, 48, 49f; embryo monocotyledon, 50, 50f; single apical cell, 62; tunica-corpus pattern, 60–61, 61f
shoot branching, 65–66
shoot buds, from roots, 125–126, 125f
shoot growth, 59–60
shoot modifications: cladophylls, 66–67; tendril, 67, 67f; thorn, 67, 67f
shoot morphology and development, 57–68; apex and apical meristem, 60–63, 60f–63f; branching, 65–66; expansion, 63–65, 64f; growth, 59–60; modifications, 66–67, 67f; reproductive shoots, 67–68
shoot system, 2, 3f; axis, 81 (*see also* stem tissues); long shoot, 57–58, 63–64, 64f; nodes and internodes, 57–60, 63–65. 81–82; short shoot, 57–58, 64, 64f
sieve areas, 76–77, 89
sieve cells, 77–78
sieve plates, 77, 77f, 89; secondary phloem, 137
sieve tube elements (or sieve elements), 76–78, 77f, 137; callose, 77, 77f; companion cells, 77f, 78; fibers, 76; P-protein, 76, 77, 89; sieve areas, 76–77, 89; sieve plates, 77, 77f, 89; sieve tubes, 76–77, 77f
sieve tubes, 76–77, 77f; secondary phloem, 137
silica bodies, 13
silique, fruit type, 54t
simple fruit, 54t, 55

simple leaf, 96, 96f, 97
siphonostele, 93, 93f
smooth endoplasmic reticulum (ER), 9
somatic embryogenesis, 51–52
sperms, 4, 5f; cells, 21, 38, 39, 39f, 40f; fertilization, 41, 42f; seed plants, 35
spherosomes, 12
spike, inflorescenece, 18f, 19
spikelet, Poaceae, 26
spiral phyllotaxy, 59f, 60, 60f
spiral tracheary elements, 74, 75f
spongy parenchyma (mesophyll), 102–103, 102f, 106; leaf expansion, 109; leaf veins, 103
sporangium (sporangia), 4, 5f; mosses, 34
spore mother cell, 5f
spores, 4, 5f. *See also* megaspore; microspore
sporogenous tissue, 20–21
sporophyte: 5, 5f; diploid, 31; embryo, 5
sporopollenin, 39; exine, 39
stamens, 20, 20f; development, 29; growth, 29–30
staminate flower, 22
starch grains/granules, 11–12, 12f
Stelaria media, embryo, 47–48, 48f
Stele: types, 92–94, 93f
stem: axillary bud, 57; internodes, 57; nodes, 57; phytomeres, 57
stem tissues, 81–94; 3-D organization and structure, 84–86, 85f; dermal tissue (epidermis), 82–83, 87–88, 82f; dicotyledons, 82–84, 82f, 84f; differentiation, 64f, 86–91, 87f; differentiation pattern, 93f; ground meristem, 64f, 86–87, 87f, 88; ground tissue, 82–83, 82f, 92f; leaf traces, 85–86, 85f; monocotyledons, 91–92, 92f; nodes and internodes, 81–82, 84–85; phloem, 82–84, 82f, 84f, 89, 90–91; procambium, 87f, 88–89, 91; promeristem, 86, 87f; protoderm, 64f, 86–88; provascular tissue, 86–87, 87f; shape, 82; stele, 92–94, 93f; structure and roles, 81–82; tracheary elements, 89; vascular tissue, primary, 82–84, 82f, 84f, 85f, 92f; xylem, 89–91

stem tissues dicotyledons: bundle cap, 82f, 83, 84; collenchyma, 83; cortex, 82, 82f, 83; cuticle, 83; epidermis, 82, 82f, 83; fibers, 83, 84; hairs, 83; hypodermis, 83; phloem, 82–84, 82f, 84f, 89, 90–91; pith, 82, 82f, 83; sclereids, 82f, 83; sclerenchyma, 83; sieve tubes, 84; vascular bundle, 82, 82f, 83–84, 84f; vascular bundle, leaf extension, 103; vascular system, 83–84

stem tissues monocotyledons: epidermis, 92f; ground tissue, 91, 92f; vascular tissues (bundles), 91, 92f

stigma, 20f, 21; pollen germination, 40, 41; self-incompatibility, 40

stipule, 96, 96f, 97

stolon, vegetative reproduction, 33, 33f

stoma (plural = stomata), 78, 79f; guard cells, 78–79, 79f; leaf blade, 79f, 101–103, 104f

storage organs, leaf, 100

storied cambium, 131–132, 131f

stratified cambium, 131–132, 131f

strawberry: runner, 33, 33f

striate (parallel) venation, 96f, 98–99, 99f

stroma, 10

stroma lamella, 10

style, 20f, 21; open canal, 41; pollen tube growth, 41; transmitting tissue, 41

styloids, 12

suberin, 120, 141; cork cells, 80

subsidiary cells, 102

succulents leaves, 106, 106f

sun leaves, 104

suspensor, 51; development, dicotyledons, 45f, 46, 46f; development, monocotyledons, 50

symmetry, bilateral, 47

sympetaly, 22

symplast, 14

synandry, 22

syncarpous ovary, 21, 26, 26f; development, 29

synergids, 21, 36–37, 37f, 38; fertilization, 41, 42f

synsepaly, 22

tannins, 12

tapetum, 20–21, 38

tap root, 112, 112f

Taraxacum (dandelion), 43

tendrils, 67, 67f, 100, 100f

tepals, 20, 24, 25f, 26

terminal branching, *Lycopodium, Selaginella*, 65

terminal cell (suspensor), 45–46, 45f

testa, 53

tetrad, spores, 5f; megapores, 35, 36f; microspores, 38, 39f

tetrarch xylem, 121, 122f

tetrasporic embryo sac (female gametophyte), 38; *Fritillaria* type, 38

thorn, 67, 67f

thylakoids, 10

tissue culture propagation, 34

tissue systems, 69–70; dermal, 69; ground (fundamental), 69, 70; stem, 81–94 (*see also* stem tissues); vascular, 69–70

tissue types: collenchyma, 71–72, 71f; cork cells, 80, 139, 139f; epidermal cells, 78–80, 79f; parenchyma, 71, 71f, 72f; sclerenchyma, 72, 73f, 83, 120; sieve elements, 76–78, 77f; tracheary elements, 73–76, 74f, 75f

tomato: flower development, 29; pollen, 39; syncarpous ovary, 29; vegetative and floral apices, 28, 28f

tonoplast, 8f, 11

tracheary elements, 15, 73–76, 74f, 75f. *See also specific types*: annular, 74, 75f; bordered pits, 74, 75f, 80; helical, 74, 75f; pitted, 74, 74f, 75f; scalariform, 74, 75f; stem tissue, 89; tracheids, 15, 74f, 76, 133–135, 134f; vessel elements, 74f, 76; xylem, 73–74, 74f, 75f, 82–84, 82f, 84f

tracheids, 15, 74f, 75f, 76; secondary xylem, 133–135, 134f

Tracheophyta, 2

transfer cells, 71

transmitting tissue, 41

triarch xylem, 121, 122f

trichomes. *See* epidermal hairs

trichosclereid, 73

trilacunar node, 86

tuber, 33, 33f

tunica, shoot apical meristem, 61, 61f

umbel, inflorescence, 18f, 19
unifacial (tube like) leaf, 109
unilacunar node, 86
uniseriate rays, 135–136, 136f, 138
unisexual plants, 22
unit membrane, 8f, 9
utricle, fruit type, 54t

vacuole, 8f, 11
vascular bundles: dicotyledon stem, 82, 82f, 83–84, 84f; monocotyledon-, stem, 91, 92f; stem to leaf extension, 84–85, 85f, 103
vascular cambium, 3f, 4, 128–133; activity, 131–133, 131f, 132f, 137f; initiation, 128–129, 128f; organization, 129–131, 130f; secondary phloem, 136–138, 137f; secondary xylem (wood), 133–136, 132f, 134f–136f; storied, 131, 131f
vascular plants, 2. *See also specific topics and types*
vascular system: leaf, 103, 104f; root, 121–122, 122f; stem, 83–86, 85f
vascular tissue, 2, 69–70; dicotyledon stem, 82, 82f, 83–84; leaf, 101–104, 99f, 102f, monocotyledon stem, 91, 91f, root, 118, 118f, 121–122, 121f, 122f; secondary body, 133–138, 132f, 134f, 135f, 137f
vegetative (asexual) reproduction, 32–34, 33f; artificial, 33–34; bulbs and corms, 33; clones, 32; natural, 32–33, 33f; regeneration, 32; rhizome, 32–33, 33f; runner, 33, 33f; stolon, 33, 33f; tubers, 33, 33f
vegetative apex, 27–28, 28f; transformation to floral or inflorescence apex, 67, 68, 28f. *See also* shoot apex

veinlets, 98, 99f
veins and vein connections, leaf, 99, 99f, 102f, 103
velamen, 119
vessel elements (or members), 74f, 76. *See also specific types*: xylem, 73–76, 74f–75f; secondary xylem, 134, 134f
Vivipary, 55

wall matrix, 13
whorled phyllotaxy (leaf arrangement), 58, 58f
wood (secondary xylem), 133–136; conifer, 133, 133f; dicotyledon, 133–136, 134f–136f; diffuse porous, 134, 135f; fusiform initials, 133; ray initials, 133; ring porous, 134, 134f

xerophytes: definition, 105; leaf form and tissues, 104f, 105–106
xylem, 2, 70, 73–76, 74f, 75f; basipetal extension, 90–91; dicotyledons, 82–84, 82f, 84f; differentiation, 88–91, 93f; fibers, 72, 74; leaf, 103, 104f; procambium, 63, 63f; root, 119, 121–122, 121f, 122f; secondary xylem (wood), 133–136, 134f–136f; stem tissue, 89–91; vascular cambium, 128–133, 128f, 130f. *See also* tracheary elements; tracheids; vessel elements
xyloglucan, 13

Zea mays (corn, maize), 22; embryo development, 50–51, 50f; root, 113; root apical meristem, 116; stem, 92f
zygote, 5f, 31; diploid, 5, 41, 42f